# Statistics and Computing

*Series Editors:*
J. Chambers
W. Eddy
W. Härdle
S. Sheather
L. Tierney

# Springer

*New York*
*Berlin*
*Heidelberg*
*Barcelona*
*Hong Kong*
*London*
*Milan*
*Paris*
*Singapore*
*Tokyo*

# Statistics and Computing

W.N. Venables
B.D. Ripley

# S Programming

With 10 Illustrations

Springer

both to interface to existing code, and to speed up methods initially written in S. Indeed, we have often used the S language for rapid prototyping, transferring key computations to compiled code in C and eventually writing a pure C version. We devote a chapter to using compiled code.

The authors may be contacted by electronic mail as

```
Bill.Venables@cmis.csiro.au
ripley@stats.ox.ac.uk
```

and would appreciate being informed of errors and improvements to the contents of this book. Errata and updates will be made available on-line (see page 4).

To avoid any confusion, S-PLUS is a commercial product, details of which may be obtained from `http://www.mathsoft.com/splus/`, and R is an Open Source project, with source and binaries available via a network of sites mirroring `http://www.r-project.org/`.

*Acknowledgements:*

This book would not be possible without the S environment which has been principally developed by Rick Becker, John Chambers and Allan Wilks; version 4 of the S language is the result of years of John Chambers' work. The S-PLUS code is the work of a much larger team acknowledged in the manuals for that system. The R project was the inspiration of Ross Ihaka and Robert Gentleman, and incorporates the work of many volunteers.

We are grateful to the many people who contributed to our understanding or have read and commented on draft material. In particular we would like to thank Rich Calaway, John Chambers, Bill Dunlap (the other authority on the S language, apart from its principal author), Nick Ellis, Stephen Kaluzny, José Pinheiro, Charles Roosen, Berwin Turlach and the R–core team (with special thanks to Kurt Hornik for help in implementing our code in R).

Bill Venables
Brian Ripley
December 1999

# Contents

# Appendices

# Typographical Conventions

Throughout this book S language constructs and commands to the operating system are set in a monospaced typewriter font `like this`. The character ~ may appear as ˜ on your keyboard, screen or printer.

We often use the prompts `$` for the operating system (it is the standard prompt for the Unix Bourne shell) and `>` for S. However, we do *not* use prompts for continuation lines, which are indicated by indentation. One reason for this is that the length of line available to use in a book column is less than that of a standard terminal window, so we have had to break lines which were not broken at the terminal.

Some of the S output has been edited. Where complete lines are omitted, these are usually indicated by

```
    . . . .
```

in listings; however most *blank* lines have been silently removed. Much of the output was generated with the options settings

```
    options(width=65, digits=5)
```

in effect, whereas the defaults are `80` and `7`.

R    R differences (at the time of writing) are often signalled by a marginal R, as here.

# Chapter 1

# Introduction

S is a language for "programming with data", in the words of the title of Chambers (1998). John Chambers of Bell Labs (formerly part of AT&T, currently part of Lucent Technologies) has been its principal designer over more than two decades, and was awarded the prestigious 1998 *Association for Computing Machinery* Award for Software Systems for, in the words of the citation,

> the S system, which has forever altered how people analyze, visualize, and manipulate data.

For the last decade it has been the major vehicle for the delivery of new statistical methodology to end users.

Most users coming to an S environment for the first time initially regard it as "another statistical package" for standard statistical analyses and graphics. In the companion volume to this one, *Modern Applied Statistics with S-PLUS* (Venables & Ripley, 1994,7,9) we have concentrated upon features of the S environments likely to be of most immediate concern to new or intermediate users while, we hope, articulating and illustrating modern statistical methodology. Reference to that work will be fairly frequent and we will use the acronym 'MASS'.[1]

However, its developers have been adamant that S is *not* a statistical language, as can be seen in the previous quote and in

> Notice that statistics is not properly part of S itself; S is a computational language and environment for data analysis and graphics.

(Becker, 1994, p. 101). Although we take a wider view of statistics than the S developers (to us "data analysis" and "data mining" are part of applied statistics), it is true that in an S environment the pre-written software is typically less comprehensive than one would expect of a statistical package. The compensating feature is that the environment itself contains a full-featured programming language in which the user can, in principle at least, implement any data analysis or graphical procedure by programming just as one would in C or FORTRAN. The libraries we wrote to enable us to do the analyses in MASS are instances of user-written extensions for the S environments. With this book we hope to encourage the reader

---

[1] Although some familiarity with MASS would be useful in motivating the material of this volume, the two books are self-contained and may be read independently.

to write (and document) similar extensions for personal use and if appropriate to contribute them to the large public archive of S extensions already available. This ready extensibility has been, we believe, the cornerstone of S's success.

We agree strongly with Chambers (1998, Chapter 1) when he contends that programming ought to be regarded as an integral part of effective and responsible data analysis. The goal of the S language, as stated in the preface of that reference and repeated frequently. is "to turn ideas into software, quickly and faithfully". We would go further and offer the view that it is the duty of the responsible data analyst to engage in this process, that is to devise, develop and implement new strategies for data analysis to meet the new challenges that constantly arise.

Turning ideas into software in this way need not be an unpleasant duty, of course: programming can be mentally very stimulating and immensely satisfying. In addition the exercise of drafting an algorithm to the level of precision that programming requires can in itself clarify ideas and promote rigorous intellectual scrutiny. In our view it is somewhat ironic that even very substantial software contributions do not seem to attract the same academic credit as refereed publications: in reality nearly every user of software becomes a more meticulous and critical reviewer than most anonymous referees!

## 1.1 Versions of S

The evolution of the S language has had four phases, characterized by books (co-)authored by Chambers and known by the predominant colours of their covers. Becker & Chambers (1984) (*brown*, with a slim supplement, Becker & Chambers, 1985) is of historical interest only. Becker *et al.* (1988) (*blue*) introduced what was known then as 'New S' and is now known as S version 2. That version had a great impact, and has laid the basis for future versions. We must note the extensive contributions of Rick Becker to the first two phases. It was the 1988 version that introduced the current extensibility, that users could write functions in the language to the same terms as the original designers. It was also serendipitous in its timing, as workstations were just becoming widespread with enough processing power and graphical capabilities to support the style of data analysis that S encourages.

Chambers & Hastie (1992) (*white*; it appeared in 1991) introduced S version 3, which built on S version 2 and introduced structures to make statistical modelling in S easier. Chambers (1998) (*green*) describes S version 4, which was a far-reaching internal redesign with a high degree of backward compatibility and a more formal object-oriented structure. Dr Chambers' current research interest is in distributed data analysis software systems[2] which is related to S development but represents a departure from the previous line.

Becker (1994) gives "a brief history of S" up to that time but concentrating on the earlier years.

---

[2]see www.omegahat.org.

The popularity of the S language grew with the availability of a commercial implementation called S-PLUS in 1988, from a start-up company in Seattle called Statistical Sciences Inc., which merged with MathSoft Inc. in 1993 and in that year acquired exclusive marketing rights to S. There has been a series of versions of S-PLUS on both UNIX and Windows, with a numbering system that is confusingly similar to those used for S versions. Versions that are likely to be still be in use are

3.x the latest versions being 3.4 (1996) on UNIX and 3.3 (1995) on Windows 3.x. This is based on S version 3.

4.0, 4.5, 2000[3] (1997–9) on Windows. Also based on S version 3. We call these collectively 4.x.

5.x (1998–) on UNIX (including Linux). Based on S version 4. Version 5.1 release 1 was used for the descriptions given here. A Windows version based on S version 4 has been announced (but not named).

To circumvent this confusion, we will refer to the 'old S engine' (S version 3; S-PLUS 3.x, 4.x) and the 'new S engine' (S version 4; S-PLUS 5.x and the future Windows version). We do not consider the Windows versions of S-PLUS 3.x further.

An Open Source system called R is being developed which is 'not unlike S'. Ihaka & Gentleman (1996), its progenitors, give some of its history: note that their title is 'not unlike' that of the brown book but R is most similar to S version 3. R is now developed in their 'spare' time by a small core team with volunteer contributions, and is available as source code and in a number of binary versions, including for Windows. In this book R differences are often signalled by a marginal R: these refer to version 0.90.1 (December 1999).

For the purposes of this book, we regard S as a language with three current implementations of different dialects. We will refer to the implementations as *engines*. One measure of the similarity of R to the old S engine is the small   R number of changes needed to run most of the examples of MASS under R (see our on-line 'R Complements' for details).

## 1.2   S programming

To be able to program in a language, one needs some acquaintance with its grammar and functionality. Probably most readers will already have a passing acquaintance from MASS, the blue book (Becker *et al.*, 1988), Spector (1994) or the S-PLUS manuals. Chapter 2 provides both a refresher course and a reference section. It is based closely on MASS, but is a little more formal and does describe many of the differences of the R implementation. However, serious programmers will want to know about the 'power features' (and, unfortunately, the pitfalls) of the language described in Chapter 3.

---

[3]which reports its version as 4.7.

All current versions of S are in some sense 'object-oriented', although this term is not used in exactly the same sense as by, say, Java or Visual Basic. Chapter 4 describes the object-orientation of the old S engine and R, and Chapter 5 the much more pervasive class-based nature of the new S engine.

To us[4] one of the most powerful ways to use S is to prototype methods in the S language and to move the time-critical parts to a compiled language and link them into the engine. That this can be done in a way that is totally transparent to the end user (who only sees a snappier response) is an outstanding feature of the development environment. We consider using compiled code in Chapter 6 and the system-specific implementations in Appendix A. It is possible in both the new S engine and in R to manipulate S objects directly in C code.

Programming in the strict sense is just one part of a software engineering project, and we consider other aspects in the last three chapters, including documentation, interfaces and efficiency.

## 1.3 On-line material

The S, C and FORTRAN code used for the examples in this book is available on-line. Point your Web browser at

```
http://www.stats.ox.ac.uk/pub/MASS3/sites.html
```

to obtain a current list of sites; please use a site near you. We expect this list to include

```
http://www.stats.ox.ac.uk/pub/MASS3/Sprog
http://www.cmis.csiro.au/S-PLUS/MASS/Sprog
http://lib.stat.cmu.edu/S/MASS3/Sprog
http://franz.stat.wisc.edu/pub/MASS3/Sprog
```

Those sites will also have documents updating this book for future changes in S-PLUS and R.

---

[4]some S programmers do disagree.

# Chapter 2

# The S Language: Syntax and Semantics

This chapter provides a reprise of the material introduced in Chapters 2 and 4 of MASS. Chapter 3 introduces more advanced language concepts that are important for programming: there are also class-oriented features which we discuss in Chapters 4 and 5.

Everything in this chapter is intended to apply equally to all the S engines unless otherwise stated. Material in this chapter not mentioned in MASS is introduced by a marginal mark ○.

## 2.1 A concise description of S objects

In this section we discuss the most important types of S objects.

### Expressions and assignments

Commands to S are either expressions or assignments. Commands are separated by either a semi-colon or a newline. The # symbol marks the rest of the line as comments.

The S prompt is > unless the command is syntactically incomplete, when the prompt changes to +.[1] The only way to extend a command over more than one line is by ensuring that *it is* syntactically correct but incomplete until the final line.

An expression command is evaluated and (normally) printed. For example

```
> 1 - pi + exp(1.7)
[1] 3.332355
```

This rule allows any object to be printed by giving its name. Note that pi is the value of $\pi$. Giving just the name of an object will normally print it or a short summary; this can be done explicitly using the function print (show in the new S engine), and summary will often give a fuller description.

---

[1]These prompts can be altered. See Section B.2 or Section A.3 in MASS.

An assignment command evaluates an expression and passes the value to a variable but the result is not printed. The recommended assignment symbol is the combination, " <- ", so an assignment in S looks like

```
a <- 6
```

which gives the object a the value 6. To improve readability of your code we strongly recommend that you put at least one space before and after binary operators, especially the assignment symbol. Assignments using the right-pointing combination " -> " are also allowed to make assignments in the opposite direction, but these are never needed and are little used in practice.

An assignment is a special case of an expression with value equal to the value assigned. When this value is itself passed by assignment to another object the result is a multiple assignment, as in

```
b <- a <- 6
```

Multiple assignments are evaluated from right to left, so in this example the value 6 is first passed to a and then to b. You will see the value of an assignment used occasionally, for example

```
if( any( nas <- is.na(x) ) ) x <- x[!nas]
```

usually to enhance the readability of the code.

Several commands can be grouped together as an expression by placing them within braces {...}. The value of a grouped expression is the value of the last command.

At the interactive level the most recently evaluated non-assignment expression in the session is stored as the variable .Last.value[2] and so may be kept as a permanent object by a following assignment such as keep <- .Last.value. This is also useful if the result of an expression is unexpectedly not printed; just use print(.Last.value).

## Naming conventions

Standard S names for objects are made up from the upper- and lower-case roman letters, the digits, 0–9, in any non-initial position and also the period, ' . ', which behaves as a letter except in names such as .37 where it acts as a decimal point. Note that S is *case sensitive*, so Alfred and alfred are distinct S names and that the underscore, _ , is *not* available as a letter in the S standard name alphabet. (Periods are often used to separate words in names, but are also used for some important conventions such as print.lm, so users should try to avoid confusion with these.)

Non-standard names are allowed using any collection of characters, but special steps have to be taken to use them. They are needed in some contexts but not

---

[2]If the expression consists of a simple name such as x, only, the .Last.value object is not changed.

in ordinary use and we suggest that they be avoided where possible. For example so-called *replacement functions* must have names that end in the two characters "<-", but such names need not be used explicitly.

In many cases[3] non-standard names may be handled by placing them within quotes. In particular this is true of function names (when used for function calls), object names to which a value is being assigned, and component names.

Avoid using system names for your own objects; in particular avoid c, q, s, t, C, D, F, I, T, diff, length, mean, pi, range, rank, time, tree and var. Using these names will probably just give a warning, but just occasionally[4] will cause a computation to fail or even give the wrong answers. There are some reserved words you will not be allowed to use, for example

```
FALSE Inf NA NaN NULL TRUE
break else for function if in next repeat return while
```

and in the S engines (but not R) F and T.

## S objects

Everything within the S language is an object, even the functions and operators. All objects have a *mode* and a *length*. The mode is what distinguishes data objects from language objects: data objects are usually of mode numeric or complex or character or logical whereas language objects are of mode function, call, expression, name, missing, ...[5]. The standard data objects are all vectors of arbitrary length: there are no explicit scalars in S.

*Lists* are used to collect together objects, for example the results of a function call, which can be of different lengths and modes. They are similar to structures in programming languages but much more dynamic, in that the number of items in the list and their names (if any) and modes are not pre-specified. The objects in lists may themselves be lists, and list objects are sometimes said to be *recursive*. The length of a list is the number of members (called components).

In contrast to recursive objects the modes numeric, complex, character, logical and NULL are known as *atomic* modes.

Objects may also have *attributes*, other objects which are attached to the main object but (unlike a list) are of a subordinate status. Examples include the names attribute of a vector and the dim attribute of an array. We may set or find all the attributes of an object by a call to attributes on the left- or right-hand side of an assignment, but this is rarely used. Rather we use the function attr to examine or set individual attributes. For example, from the S-PLUS version of our library MASS,

---

[3] one exception is formal argument names in R and in the old S engine.
[4] see page 154 for one example.
[5] For a full list in S-PLUS see ?mode on your system.

```
> attributes(mdeaths)
$class:
[1] "rts"
$tspar:
 start     deltat frequency
  1974 0.08333333        12
attr($tspar, "units"):
[1] "months"

> attr(mdeaths, "tspar")["start"] <- 1975
> attr(mdeaths, "tspar")
 start     deltat frequency
  1975 0.08333333        12
attr(, "units"):
[1] "months"
```

Whole objects can be created including attributes with the function `structure`, for example

```
res <- structure(x.p, SE = SE, p = p)
```

creates an object `res` with attributes `SE` and `p`. (There is another example on page 29.)

In most cases attributes may be assigned to an object using special *replacement* functions (to be discussed later on page 10). Very often these do some consistency checking as well as set the attribute, so where they exist they should be used in preference to using `attr` on the left side of an assignment. Thus although a `"names"` attribute may be set with the `attr` function it is both clearer and safer to do so with the `names` replacement function.

S objects may also have a class, and in the new S engine they always do. The old-style classes are stored as attribute `class`: they can affect markedly the way the object is handled. (So can other attributes; the presence of a `dim` attribute signals that the object is an array or matrix.) Classes are a large topic which we discuss in detail in Chapters 4 and 5.

Objects also have a *storage mode* which S uses to differentiate between integer, real and double precision objects, and can be queried or set by the function `storage.mode`. It is almost exclusively used in connection with compiled code (see page 123), but occasionally the storage mode of an object will be set to R    `"single"` (not R[6]) or `"integer"` to reduce its size. R objects have another way to give information using the function `typeof`, which often gives the same information as `storage.mode`.

The object NULL which represents 'nothing' is rather special: it has its own mode and storage mode (both `"NULL"`). Its length is zero, but it is distinct from other zero-length objects.

---

[6]R does not use single-precision variables internally, but setting the `storage.mode` to `"single"` sets an attribute which ensures that suitable copies are made when calling compiled code.

## Functions

S is a functional language, and we have already seen informally many examples of S functions. Functions in S are themselves S objects and a powerful feature is that they can be manipulated as S objects (Section 3.5).

When functions are called their arguments may be given names or determined by their order (the complete rules are given in Section 3.1). Generally the first few arguments are not named, but it can be helpful (or necessary) to name later ones, as in

```
res <- width.SJ(galaxies, method = "dpi")
```

We could have written this as

```
res <- width.SJ(galaxies, , , , "dpi")
```

but this is prone to error and vulnerable to changes in the definition of the function.

It is usually possible to abbreviate the argument names, but this is best avoided in programming.

## Vectors and matrices

Standard vectors consist of numeric or logical values or character strings (and they may not be mixed). Normally it is unnecessary to be aware whether the numeric values are integer, real or even complex, or whether they are stored to single or double precision; S handles all the possibilities in a unified way. For the record, there are six basic types of vectors

```
logical   integer   double   single (not R)
complex   character
```

and lists are also vectors, as are some language objects (Section 3.5).

The simplest way to create a vector is to specify its elements by the function c (for concatenate):

```
> mydata <- c(2.9, 3.4, 3.4, 3.7, 3.7, 2.8, 2.8, 2.5, 2.4, 2.4)
> colours <- c("red", "green", "blue", "white", "black")
```

Character strings may be entered with either double or single quotes (in matching pairs), but will always be printed with double quotes. Note that a character vector is a vector of character strings, not of single characters.

Logical values are represented as T and F :

```
> mydata > 3
 [1] F T T T T F F F F F
```

and may also be entered as TRUE and FALSE. In R, the roles are reversed; TRUE and FALSE are the logical values and T and F are variables with those values, initially.

Complex values are entered in the form 3.2 + 2.2i; use x + 0i if the value of x is real but is to be handled in the complex number field.

The elements of a vector can also be named and accessed by name:

```
> names(mydata) <- c('a','b','c','d','e','f','g','h','i','j')
> mydata
  a   b   c   d   e   f   g   h   i   j
2.9 3.4 3.4 3.7 3.7 2.8 2.8 2.5 2.4 2.4
> names(mydata)
 [1] "a" "b" "c" "d" "e" "f" "g" "h" "i" "j"
> mydata["e"]
  e
3.7
```

An assignment with the left-hand side something other than a simple identifier, as in the first of these, will be called a *replacement*. They are disguised calls to a special kind of function called a *replacement function* that re-constructs the entire modified object. The expanded call in this case[7] has the form

```
mydata <- "names<-"(mydata,
              c('a','b','c','d','e','f','g','h','i','j'))
```

Note that in the third command above the function `names` *extracts* the names of the components, and so is a completely different function from the replacement function of (apparently) the same name used in the first line.

More generally, several elements of a vector can be selected by giving a vector of element names or numbers or a logical vector:

```
> letters[1:5]
[1] "a" "b" "c" "d" "e"
> mydata[letters[1:5]]
  a   b   c   d   e
2.9 3.4 3.4 3.7 3.7
> mydata[mydata > 3]
  b   c   d   e
3.4 3.4 3.7 3.7
```

(The index `1:5` is a quick way of giving the regular sequence `c(1,2,3,4,5)` ; see page 19) We will refer to taking *subsets* of vectors (although to mathematicians they are subsequences), and discuss this more fully in Section 2.3.

*Matrices*

Giving a vector a `dim` attribute allows, and sometimes causes, it to be treated as a matrix or array. For example

```
> names(mydata) <- NULL      # remove the names
> dim(mydata) <- c(2, 5)
> mydata
     [,1] [,2] [,3] [,4] [,5]
[1,]  2.9  3.4  3.7  2.8  2.4
[2,]  3.4  3.7  2.8  2.5  2.4
> dim(mydata) <- NULL
```

---

[7]This will not work as a direct call in the old S engine.

Notice how the matrix is filled down columns rather than across rows. The final assignment removes the dim attribute and restores mydata to a vector.[8] A simpler way to create a matrix from a vector is to use matrix

```
> matrix(mydata, 2, 5)
     [,1] [,2] [,3] [,4] [,5]
[1,]  2.9  3.4  3.7  2.8  2.4
[2,]  3.4  3.7  2.8  2.5  2.4
```

The function matrix can also fill matrices by row,

```
> matrix(mydata, 2, 5, byrow = T)
     [,1] [,2] [,3] [,4] [,5]
[1,]  2.9  3.4  3.4  3.7  3.7
[2,]  2.8  2.8  2.5  2.4  2.4
```

As these displays suggest, matrix elements can be accessed as mat[m,n], and whole rows and columns by mat[m,] and mat[,n] respectively. In the new S engine such objects are of class "matrix".

Arrays are multi-way extensions of matrices, formed by giving a dim attribute of length three or more. They can be accessed by A[r, s, t] and so on. The function array (page 25) is a convenient way to generate arrays. The dimensions can be given names: see dimnames on page 25.

The function dim can be used to find the dimensions of an array or matrix. For matrices two convenience functions nrow and ncol are provided[9] to access the number of rows and columns respectively.

The functions cbind and rbind can be used to join together two or more vectors or matrices, column-wise and row-wise respectively.

*Creating vectors*

The functions numeric, character and logical generate a vector of length specified by their first argument. They are useful to allocate storage which can be filled by assigning to parts of the vector, and this uses memory more efficiently than extending an existing vector. More generally, the function vector generates a vector of specified mode and length.

Note that the length of a vector can be zero, which is occasionally useful. The expression numeric(0) is both the expression to create an empty numeric vector and the way it is represented when printed. It has length zero. It may be described as "a vector such that if there were any elements in it, they would be numbers"! Note that this is quite distinct from NULL; a zero-length vector has a mode and storage mode and should be thought of as an empty container, not as no container.

R also allows zero for one or more of the dimensions of a matrix or array.     R

---

[8]and in the new S engine this removes the names.
[9]Note that the names are singular: it is all too easy to write nrows!

## Lists

A list is used to collect together items of different types. For example, an employee record might be created by

```
Empl <- list(employee = "Anna", spouse = "Fred", children = 3,
             child.ages = c(4,7,9))
```

As this example shows, the elements of a list do not have to be of the same length. The components of a list are always numbered and may always be referred to as such. Thus `Empl` is a list of length 4, and the individual components may be referred to as `Empl[[1]]`, `Empl[[2]]`, `Empl[[3]]` and `Empl[[4]]`. Further, since `Empl[[4]]` is a vector, `Empl[[4]][1]` is its first entry. However, it is more convenient to refer to the components explicitly by name, in the form

```
> Empl$employee
[1] "Anna"
> Empl$child.ages[2]
[1] 7
```

Names of components may be abbreviated to the minimum number of letters needed to identify them uniquely. Thus `Empl$employee` may be minimally specified as `Empl$e` since it is the only component whose name begins with the letter 'e', but `Empl$children` must be specified as at least `Empl$childr` because of the presence of another component called `Empl$child.ages`. However, abbreviating names is usually not a good idea, especially in programming, as adding or removing components later can give unexpected results.

In many respects lists are like vectors. For example, the vector of component names is simply a `names` attribute of the list like any other object and may be treated as such; to change the component names of `Empl` to `a`, `b`, `c` and `d` we can use the replacement

```
> names(Empl) <- letters[1:4]
> Empl[3:4]
$c:
[1] 3
$d:
[1] 4 7 9
```

Notice that we select *components* as if this were a vector (and not `[[3:4]]` as might have been expected). (The distinction is that `[ ]` returns a list with the selected component(s) so `Empl[3]` is a list with one component, whereas `Empl[[3]]` extracts or replaces that component only.)

The concatenate function, `c`, can also be used to concatenate lists or to add components, so

```
Empl <- c(Empl, service = 8)
```

would add a component for years of service. This has a named argument `recursive`; if this is true the list arguments are unlisted before being joined together. Thus

```
c(list(x = 1:3, a = 3:6), list(y = 8:23, b = c(3, 8, 39)))
```

is a list with four (vector) components, but adding `recursive=T` gives a vector of length 26. (Try both to see.)

The function `unlist` converts a list to a vector:

```
> unlist(Empl)
 employee spouse children child.ages1 child.ages2 child.ages3
 "Anna"   "Fred" "3"      "4"         "7"         "9"
> unlist(Empl, use.names = F)
[1] "Anna" "Fred" "3"     "4"     "7"     "9"
```

which can be useful for a compact printout (as here). (Mixed types will all be converted to character, giving a character vector.)

Some care is needed when handling `NULL` values in lists. For example,

```
Empl <- list(employee="Anna", spouse="Fred", children=3,
             child.ages=c(4,7,9))
Empl["spouse"] <- NULL
```

removes the `spouse` component in R but does nothing in S, whereas                R

```
Empl["spouse"] <- list(NULL)
```

sets the `spouse` component to `NULL` in both R and S, and

```
Empl[["spouse"]] <- NULL
```

removes it in both.

The set of components of a list is dynamic, as elements can be added or deleted. This can cause difficulties with partial matching of component names, which is thus best avoided. To do so, use numbers rather than names to index lists, and as the numbering is also dynamic, use function `match` to find the current number. Thus

```
Empl[[match("spouse", names(Empl))]]
```

is guaranteed to extract the `spouse` component of the list, or return `NULL` if there is not one. If we need to distinguish between not having a component and having a component with value `NULL`, we need to examine the result of the match directly.

The function `vector` can also be applied to mode `"list"`;

```
xx <- vector("list", 6)
names(xx) <- letters[1:6]
```

generates a list, names the components `"a"` to `"f"` and initializes them to be `NULL`.

## Factors

A factor is a special type of vector, normally used to hold a categorical variable, for example

```
> citizen <- factor(c("uk","us","no","au","uk","us","us"))
> citizen
[1] uk us no au uk us us
```

Although this is entered as a character vector, it is printed without quotes. Appearances here are deceptive, and a special `print` method is used. Internally the factor is stored as a set of codes, and an attribute giving the *levels*:

```
> print.default(citizen)
[1] 3 4 2 1 3 4 4
attr(, "levels"):
[1] "au" "no" "uk" "us"
attr(, "class"):
[1] "factor"
> unclass(citizen)   # oldUnclass(citizen) for the new S engine
[1] 3 4 2 1 3 4 4
```

Using a factor indicates to many of the statistical functions that this is a categorical variable (rather than just a vector of labels), and so it is treated specially.

By default the levels are sorted into alphabetical order, and the codes assigned accordingly. Some of the statistical functions give the first level a special status, so it may be necessary to specify the levels explicitly:

```
> citizen <- factor(c("uk","us","no","au","uk","us","us"),
      levels = c("us", "fr", "no", "au", "uk"))
> citizen
[1] uk us no au uk us us
Levels:
[1] "us" "fr" "no" "au" "uk"
```

Note that levels which do not occur can be specified, in which case the levels *are* printed. This often occurs when subsetting factors.[10] Our function `relevel` (page 82) provides a convenient way to change the base level of a factor.

Sometimes the levels of a categorical variable are naturally ordered, as in

```
> income <- ordered(c("Mid","Hi","Lo","Mid","Lo","Hi","Lo"))
> income
[1] Mid Hi  Lo  Mid Lo  Hi  Lo

 Hi < Lo < Mid
> as.numeric(income)
[1] 3 1 2 3 2 1 2
```

---

[10]An extra argument may be included when subsetting factors, as in `f[i, drop=T]`, to include only those levels that occur in the subset. Under the default, `drop=F`, the levels are not changed.

Again the effect of alphabetic ordering is not what is required, and we need to set the levels explicitly:

```
> inc <- ordered(c("Mid","Hi","Lo","Mid","Lo","Hi","Lo"),
    levels = c("Lo", "Mid", "Hi"))
> inc
[1] Mid Hi  Lo  Mid Lo  Hi  Lo

 Lo < Mid < Hi
```

Ordered factors are a special case of factors that some functions (including print) treat in a special way.

One point to watch is that if it is necessary to convert a factor with numeric levels to a numeric vector, this must be done by

```
as.numeric(as.character(x))
```

which will convert any non-numeric values to NAs (see page 21). If the internal codes are required, use as.numeric or, better, unclass (oldUnclass for the new S engine) and *not* the function codes (which re-codes).

## Data frames

A data frame is the type of object normally used in S to store a data matrix. It should be thought of as a list of variables of the same length, but possibly of different types (numeric, character or logical). Consider our data frame painters:

```
> painters
            Composition Drawing Colour Expression School
  Da Udine           10       8     16          3      A
  Da Vinci           15      16      4         14      A
Del Piombo            8      13     16          7      A
Del Sarto            12      16      9          8      A
Fr. Penni             0      15      8          0      A
  ....
```

which has four numerical variables and one character variable. Since it is a data frame, it is printed in a special way. The components are printed as columns (rather than as rows as vector components of lists are) and there is a set of names, the row.names, common to all variables.

```
> row.names(painters)
[1] "Da Udine"        "Da Vinci"         "Del Piombo"
[4] "Del Sarto"       "Fr. Penni"        "Guilio Romano"
[7] "Michelangelo"    "Perino del Vaga"  "Perugino"
  ....
```

Further, neither the row names nor the values of the character variable appear in quotes.

Data frames can be indexed in the same way as matrices:

```
> painters[1:5, c(2, 4)]
           Drawing Expression
  Da Udine       8          3
  Da Vinci      16         14
Del Piombo      13          7
 Del Sarto      16          8
 Fr. Penni      15          0
```

But they may also be indexed as lists, which they are. Note that a single index behaves as it would for a list, so `painters[c(2,4)]` gives a data frame of the second and fourth variables which is the same as `painters[, c(2,4)]`.

Variables that satisfy suitable restrictions (having the same length, and the same names, if any) can be collected into a data frame by the function `data.frame`, which resembles `list`:

```
mydat <- data.frame(MPG, Dist, Climb, Day = day)
```

(although data frames are most commonly created by `read.table`.)

There is a side effect of `data.frame` that needs to be considered; all character and logical[11] columns are converted to factors unless their names are included in `I()` so, for example,

```
mydat <- data.frame(MPG, Dist, Climb, Day = I(day))
```

preserves `day` as a character vector, `Day`.

Compatible data frames can be joined by `cbind`, which adds columns of the same length, and `rbind`, which stacks data frames vertically. The result is a data frame with appropriate names and row names.

It is also possible to include matrices and lists within data frames. If a matrix[12] is supplied to `data.frame`, it is as if its columns were supplied individually; suitable labels are concocted. If a list is supplied, it is treated as if its components had been supplied individually.

## Coercion

There is a series of functions named `as.xxx` that convert to the specified type in the best way possible. For example, `as.matrix` will convert a numerical data frame to a numerical matrix, and a data frame with any character or factor columns to a character matrix. The function `as.character` is often useful to generate names and other labels.

Functions `is.xxx` test if their argument is of the required type. Note that these do not always behave as one might guess; for example applying `is.vector` to `mydata` (page 10) will be *false* in the S engines (but not R) as this tests for a 'pure' vector without any attributes such as names. Similarly, `as.vector` has the (often useful) side effect of discarding all attributes.

The new S engine has general functions `is` and `as` discussed on page 105.

---

[11] logical columns are not converted in the new S engine.

[12] matrices of class `model.matrix` are an exception: they are included as a single object.

## 2.2 Arithmetical expressions

We have seen that a basic unit in S is a vector. Arithmetic operations are performed on vectors, element by element. The standard operators + - * / ^ are available, where ^ is the power (or exponentiation) operator (so x^y gives $x^y$ ). Note that operators in S are just functions, so we could write 2+3 as "+"(2,3).

Numeric vectors can be complex, and almost all the rules for arithmetical expressions apply equally to complex quantities. Functions Re and Im return the real and imaginary parts. Note that complex arithmetic is not used unless explicitly requested, so sqrt(x) for x real and negative produces an error. If the complex square root is desired use sqrt(as.complex(x)) or sqrt(x + 0i).

### Operator precedence

The formal precedence of operators is given in Table 2.1. However, as usual it is better to use parentheses to group expressions rather than rely on remembering these rules. They can be found on-line in S-PLUS by help(Syntax).

---

**Table 2.1**: Precedence of operators, from highest to lowest.

| | |
|---|---|
| $ | for list extraction |
| @ | (new S engine) for slot extraction |
| [ [[ | vector and list element extraction |
| ^ | exponentiation |
| - | unary minus |
| : | sequence generation |
| %% %/% %*% | and other special operators %...% |
| * / | multiply and divide |
| + - ? | addition, subtraction, documentation |
| < > <= >= == != | comparison operators |
| ! | logical negation |
| & \| && \|\| | logical operators |
| ~ | formula |
| <<- | assignment within a function (see page 62) |
| <- _ -> = (new S engine) | assignment |

---

### The recycling rule

The expression y + 2 is a syntactically natural way to add 2 to each element of the vector y, but 2 is a vector of length 1 and y may be a vector of any length. A convention is needed to handle vectors occurring in the same expression but not all of the same length. The value of the expression is a vector with the same length as that of the longest vector occurring in the expression. Shorter vectors are *recycled* as often as need be until they match the length of the longest vector. In particular a single number is repeated the appropriate number of times. Hence

```
x <- c(10.4, 5.6, 3.1, 6.4, 21.7)
y <- c(x, x)
v <- 2 * x + y + 1
```

generates a new vector v of length 10 constructed by

1. repeating the number 2 five times to match the length of the vector x and multiplying element by element, and

2. adding together, element by element, 2*x repeated twice, y as it stands, and 1 repeated ten times.

Fractional recycling is allowed in the old S engine and in R, with a warning, but in the new S engine it is an error.

## Some standard functions

1. There are several functions to convert to integers; round will normally be preferred, and rounds to the nearest integer. (It can also round to any number of digits in the form round(x, 3). Using a negative number rounds to a power of ten, so that round(x,-3) rounds to thousands.) Each of trunc, floor and ceiling round in a fixed direction, towards zero, down and up respectively. round is documented to round to even, so round(2.5) is 2.

2. Other arithmetical operators are %/% for integer divide and %% for modulo reduction.[13]

3. The common functions are available, including abs, sign, log, log10, sqrt, exp, sin, cos, tan, acos, asin, atan, cosh, sinh and tanh with their usual meanings. Note that the value of each of these is a vector of the same length as its argument. In R and the old S engine function log has a second argument, the base of the logarithms, defaulting to *e*. However, in the new S engine, log has only one argument, and logb must be used for 'log to base'.

   Less common functions are gamma and lgamma ($\log_e \Gamma(x)$).

4. There are functions sum and prod to form the sum and product of a whole vector, as well as cumulative versions cumsum and cumprod.

5. The functions max(x) and min(x) select the largest and smallest elements of a vector x. The functions cummax and cummin give cumulative maxima and minima.

---

[13]The result of e1 %/% e2 is floor(e1/e2) if e2!=0 and 0 (S) or NA (R) if e2==0. The result of e1 %% e2 is e1-floor(e1/e2)*e2 if e2!=0 and e1 otherwise (see Knuth, 1968, §1.2.4). Thus %/% and %% always satisfy e1==(e1%/%e2)*e2+e1%%e2. These rules are not followed for values of mode / class "integer" in the current S engines.

6. The functions pmax(x1, x2, ...) and pmin(x1, x2, ...) take an arbitrary number of vector arguments and return the element-by-element maximum or minimum values, respectively. Thus the result is a vector of length that of the longest argument and the recycling rule is used for shorter arguments.

7. The function range(x) returns c(min(x), max(x)) for a vector x. If range, max or min is given several arguments these are first concatenated into a single vector.

8. sort returns a vector of the same size as x with the elements arranged in increasing order. A second argument to sort allows partial sorting. Often more usefully the functions sort.list and order produce an index vector which will arrange their leading argument in increasing order. A second argument to sort.list allows partial sorting whereas additional arguments to order allow tie breaking.

9. The function rev arranges the components of a vector or list in reverse order. duplicated produces a logical vector with value T only where a value in its vector argument has occurred previously and unique removes such duplicated values.

10. Functions union, intersect, setdiff and is.element [14] enact the set operations $A \cup B$, $A \cap B$, $A \cap \overline{B}$ and $x \in A$ (elementwise). Their arguments (and hence values) may be vectors of any mode but, like true sets, they should contain no duplicated values.

### Generating regular sequences

There are several ways in S to generate sequences of numbers. For example, 1:30 is the vector c(1, 2, ..., 29, 30). A construction such as 10:1 may be used to generate a sequence in reverse order.

The function seq is a more general facility for generating sequences. It has five arguments, only some of which may be specified in any one call. The first two arguments, named from and to, if given specify the beginning and end of the sequence, and if these are the only two arguments the result is the same as the colon operator. That is seq(2,10) and seq(from=2, to=10) give the same vector as 2:10. The third and fourth arguments to seq are named by and length, and specify a step size and a length for the sequence. If by is not given, the default by=1 is used. For example,

```
s3 <- seq(-5, 5, by=0.2)
s4 <- seq(length=51, from=-5, by=0.2)
```

generate in both s3 and s4 the vector $(-5.0, -4.8, -4.6, \ldots, 4.6, 4.8, 5.0)$.

The fifth argument is named along and has a vector as its value. If it is the only argument given it creates a sequence 1, 2, ..., length(*vector*), or the

---

[14] not in S-PLUS 3.x.

empty sequence if the value is empty. (This makes `seq(along=x)` preferable to `1:length(x)` in most circumstances.) If specified rather than `to` or `length` its length determines the length of the result.

Finally, if `seq` is only given one *unnamed* argument it is taken as an `along` argument if its length is unequal to one, but as a `to` argument if its length is one with `from=1` assumed. Thus `seq(6)` is also equivalent to `1:6` but `seq(x)` is equivalent to `seq(along=x)` if `length(x)` is not equal to one. Such shortcuts are convenient in interactive use, but pitfalls for programmers.

A companion function is `rep` which can be used to repeat an object in various ways. The simplest form is

```
s5 <- rep(x, times=5)
```

which will put five copies of `x` end-to-end in `s5`. A `times=v` argument may specify a vector of the same length as the first argument, `x`. In this case the elements of `v` must be non-negative integers, and the result is a vector obtained by repeating each element in `x` a number of times as specified by the corresponding element of `v`. Some examples will make the process clear:

```
x <- 1:4         # puts c(1,2,3,4)                    into x
i <- rep(2, 4)   # puts c(2,2,2,2)                    into i
y <- rep(x, 2)   # puts c(1,2,3,4,1,2,3,4)            into y
z <- rep(x, i)   # puts c(1,1,2,2,3,3,4,4)            into z
w <- rep(x, x)   # puts c(1,2,2,3,3,3,4,4,4,4)        into w
```

## Logical expressions

Logical vectors are most often generated by *conditions*. The logical operators are `<`, `<=`, `>`, `>=` (which have self-evident meanings), `==` for exact equality and `!=` for exact inequality. If `c1` and `c2` are vector valued logical expressions, `c1 & c2` is their intersection ('and'), `c1 | c2` is their union ('or') and `!c1` is the negation of `c1`. These operations are performed separately on each component with the recycling rule applying for short arguments.

Logical vectors may be used in ordinary arithmetic. They are *coerced* into numeric vectors, false values becoming `0` and true values becoming `1`. For example, assuming the value or values in `sd` are positive

```
N.extreme <- sum(y < ybar - 3*sd | y > ybar + 3*sd)
```

would count the number of elements in `y` that were farther than `3*sd` from `ybar` on either side. The right-hand side can be expressed more concisely as `sum(abs(y-ybar) > 3*sd)`.

The function `xor` computes (element-wise) the exclusive or of its two arguments.

The functions `any` and `all` are useful to collapse a logical vector. The function `all.equal` provides a way to test for equality up to a tolerance if appropriate.

Note that in logical expressions factors are implicitly coerced to character. Thus if `f` is a factor, `f == levels(f)[1]` will identify elements from the first category. Furthermore in an expression such as `f == 1` the right hand side will be coerced to character as well, becoming equivalent to `f == "1"`.

In the new S engine the function `identical` tests if two objects are identical, for example

```
> x <- rnorm(101)
> all.equal(x, x+.Machine$double.eps)
[1] T
> identical(x, x+.Machine$double.eps)
[1] F
```

## Missing, indefinite and infinite values

In this section we discuss missing and special values of a vector; missing values of function arguments are considered in Section 3.2.

### *The missing value marker,* `NA`

Not all the elements of a vector may be known. When an element or value is 'not available' or a 'missing value', a place within a vector may be reserved for it by assigning the special value `NA`. All modes of data vectors (numeric, complex, character, logical) can have missing values.

In general any operation on an `NA` becomes an `NA`. The motivation for this rule is simply that if the specification of an operation is incomplete, the result cannot be known and hence is not available. Thus the mean of a vector containing one or missing values is missing.

The function `is.na(x)` gives a logical vector of the same length as `x` with values that are true if and only if the corresponding element in `x` is `NA`.

```
ind <- is.na(z)
```

Notice that the logical expression `x == NA` is not equivalent to `is.na(x)`. Since `NA` is really not a value but a marker for a quantity that is not available, the first expression is incomplete. Thus `x == NA` is a vector of the same length as `x` *all* of whose values are `NA` irrespective of the elements of `x`.

The preferred way to set a value to missing in the new S engine to use `is.na` on the left-hand side of an assignment, as in

```
> is.na(mydata)[6] <- T      # new S engine
> mydata
 [1] 2.9 3.4 3.4 3.7 3.7  NA 2.8 2.5 2.4 2.4
```

S functions differ markedly in their policy for handling missing values. Some have an option `na.rm=T` to remove missing values. Many will omit rows that contain a missing value after reducing the data matrix to only those columns needed for a calculation. For statistical functions the `na.action` argument may allow other possibilities. The default na.action is usually na.fail which causes the procedure to stop; the alternative na.omit implements the row-omission policy.

Missing values are output as NA, and can be input as NA or by ensuring that a value is missing (for example, that a field is blank). When character strings are coerced to mode numeric the result is NA unless the string parses as a number.

It is occasionally useful to know that x <- NA sets x to a *logical* vector, not a numeric vector as one might expect.

## IEEE arithmetic

Most of the systems on which S can be run use the IEEE 754 conventions for extended arithmetic. This allows the numeric values Inf, -Inf and NaN to be used as floating-point (and perhaps integer) numbers. Here NaN stands for 'not a number', and is the result of operations such as 0./0. and Inf - Inf whose value is undefined. These can be tested for by the functions is.finite, is.infinite, is.nan and is.number (not in R).

Note that the value NaN is distinct from the missing value NA, but the distinction made in S is not a clean one, hence the need for several test functions. The following table shows the results of the tests:

|             | 2.0 | Inf | -Inf | NaN | NA |
|-------------|-----|-----|------|-----|----|
| is.na       | F   | F   | F    | T   | T  |
| is.finite   | T   | F   | F    | F   | F  |
| is.infinite | F   | T   | T    | F   | F  |
| is.nan      | F   | F   | F    | T   | F  |
| is.number   | T   | T   | T    | F   | F  |

Note that is.number is the opposite of is.na, and that a NaN is regarded as a missing value by these two functions. In the S engines a value of NaN is printed as NA; in R it is printed as NaN. To emphasize this, look at

```
> x <- NaN
> x   # in S-PLUS, but not R
[1] NA
```

Even more confusingly, this also happens in format, but not as.character and cat (which give NaN), so (by implicit coercion of x to character)

```
> x == "NA"
[1] F
```

Under some circumstances missing values in datasets imported into S-PLUS are represented as NaN rather than NA; this can cause confusion when compiled code is used or where the author of the analysis code has been careful to distinguish them.

A further cause of confusion can be factors with levels that are numbers plus NA or NaN; these may print like numeric vectors but the functions is.na and so on do not apply to them.

## 2.3 Indexing

We have already seen how subsets of the elements of a vector (or an expression evaluating to a vector) may be selected by appending an *index vector*, in square brackets, to the name of the vector. For vector objects, index vectors can be any of five distinct types:

1. **A logical vector.** The index vector must be of the same length as the vector from which elements are to be selected. Values corresponding to true in the index vector are selected and those corresponding to false omitted. For example,

   ```
   z <- (x+y)[!is.na(x) & x > 0]
   ```

   creates an object z and places in it the values of the vector x+y for which the corresponding value in x was positive (and non-missing).

2. **A vector of positive integers or a factor.** In this case the values in the index vector must lie in the set { 1, 2, ..., length(x) }. The corresponding elements of the vector are selected and concatenated, in that order, in the result. The index vector can be of any length and the result is of the same length as the index vector. For example x[6] is the sixth component of x and x[1:10] selects the first 10 elements of x (assuming length(x) ⩾ 10, otherwise there will be an error).

   Note that as index vectors, factors behave as the numeric vectors of codes. Thus if f is a factor, levels(f)[f] is equivalent to as.character(f).

3. **A vector of negative integers.** This specifies the values to be *excluded* rather than included. Thus

   ```
   y <- x[-(1:5)]
   ```

   drops the first five elements of x.

4. **A vector of character strings.** This possibility only applies where an object has names to identify its components. In that case a subvector of the names vector may be used in the same way as the positive integers in case **2.** For example,

   ```
   > longitude <- state.center[["x"]]
   > names(longitude) <- state.name
   > longitude[c("Hawaii", "Alaska")]
     Hawaii  Alaska
   -126.25 -127.25
   ```

   finds the longitude of the geographic centres of the two most western states of the USA. The names are retained in the result.

5. **The index position may be empty.** In this case all components are selected and thus the behaviour is as if the index vector were 1:length(object). For example the two assignments

   ```
   x[ ] <- 0
   x <- 0
   ```

are in principle radically different. For the first assignment to be valid at all x must already exist as a vector object.[15] All components of x are then replaced by 0 but attributes such as length and names remain unchanged. The second assignment replaces (or masks) any existing object called x by a new numeric object of length 1 and value 0.

As we have seen in this last example, a vector with an index expression attached can also appear on the left-hand side of an assignment, making the operation a *replacement*. In this case the assignment operation appears to be performed only on those elements of the vector implied by the index. For example,

```
x[is.na(x)] <- 0
```

replaces any missing values in x by zeros. Note that this is really a disguised call to the function "[<-", which copies[16] the entire object.

The case of a zero index falls outside these rules. A zero index in a vector of an expression being assigned passes nothing, and a zero index in a vector to which something is being assigned accepts nothing. For example

```
> a <- 1:4; a[0]
numeric(0)
> a[0] <- 10; a
[1] 1 2 3 4
```

Zero indices may be included with otherwise negative indices or with otherwise positive indices but not with both positive and negative.

Non-integer numeric indices are truncated (towards zero) to an integer value.

Another case to be considered is if the absolute value of an index falls outside the range 1, ..., length(x). In an expression this gives NA if positive and R imposes no restriction[17] if negative. On the left-hand side of a replacement, a positive index greater than length(x) extends the vector, assigning NAs to any gap, and a negative index less than -length(x) is ignored.

## Array indices

An *array* can be considered as a multiply indexed collection of data entries. Any array with just two indices is called a *matrix*, and this special case is perhaps the most important.

A *dimension vector* is a vector of positive integers of length at least 1. If its length is $k$ then the array is $k$-dimensional, or as we prefer to say, $k$-indexed. The values in the dimension vector give the upper limits for each of the $k$ indices. Note that singly subscripted arrays do not exist other than as vectors.

A vector can be used by S as an array only if it has a dimension vector as its dim attribute. Suppose, for example, a is a vector of 150 elements. Either of the assignments

---

[15]In the new S engine it must also reside in the local database.

[16]in principle. In the new S engine it may create references to parts of the original object.

[17]in the S engines; it gives an error in R.

```
a <- array(a, dim=c(3,5,10)) # make a a 3x5x10 array
dim(a) <- c(3,5,10)          # alternative direct form
```

gives it the `dim` attribute that allows it to be treated as a $3 \times 5 \times 10$ array. The elements of a may now be referred to either with one index, as before, as in `a[!is.na(z)]` *or* with three comma-separated indices, as in `a[2,1,5]`.

To create a matrix the function `array` may be used, or the simpler function `matrix`. For example, to create a $10 \times 10$ matrix of zeros we could use

```
Zmat <- matrix(0, nrow=10, ncol=10)
```

Note that a vector used to define an array or matrix is recycled if necessary, so the 0 here is repeated 100 times.

It is important to know how the two indexing conventions correspond; which element of the vector is `a[2,1,5]` ? S arrays use 'column-major order' (as used by FORTRAN but not C). This means the first index moves fastest, and the last slowest. The function `matrix` has an additional argument `byrow` which if set to true allows a matrix to be generated from a vector in row-major order.

Each of the dimensions can be given a set of names. The names are stored in the `dimnames` attribute which is a list of (possibly NULL) vectors of character strings. For example, we can name the first two dimensions of a by

```
dimnames(a) <- list(d1 = letters[1:3],
                    d2 = c("i", "ii", "iii", "iv", "v"),
                    d3 = NULL)
```

Note that the dimensions themselves can be named by giving names to the components of the `dimnames` list.

For a $k$-fold indexed array any of the five forms of indexing is allowed in each index position. In the case of an empty index the implied range is the full range for that index position. So if a is a $3 \times 5 \times 10$ array, then `a[1:2,,]` is the $2 \times 5 \times 10$ array obtained by omitting the last level of the first index. The same sub-array could be specified in this case by `a[-3,,]`. There is one additional indexing possibility for arrays:

**6.** An array may be indexed by a matrix.   In this case if the array is $k$-indexed the index matrix must be an $m \times k$ matrix with integer entries and each row of the matrix is used as a set of indices specifying one element of the array. Thus the matrix specifies $m$ elements of the array to be extracted or replaced.

To give a simple example of a matrix index, consider extracting the antidiagonal elements of a square matrix X, that is, `X[1,n]`, `X[2,n-1]`, ..., `X[n,1]` as a vector, say Xad, and then setting these elements to zero.

```
n <- dim(X)[1]                   # same as nrow(X)
ad <- matrix(c(1:n, n:1), n, 2)
Xad <- X[ad]                     # extract the antidiagonal
X[ad] <- 0                       # replace by zero.
```

(There is a function `diag` which can be used for both operations on the usual diagonal.)

There are convenience functions `row` and `col` that can be applied to matrices to produce a matrix of the same size filled with the row or column number. Thus to extract the upper triangle of a square matrix `A` we can use

```
A[col(A) >= row(A)]
```

This is a logical vector index, and so returns the upper triangle in column-major order. For the lower triangle we can use `<=` or the function `lower.tri`. A few S functions want the lower triangle of a symmetric matrix in row-major order: note that this is the upper triangle in column-major order. As an example of this sort of manipulation we can convert between a symmetric array of distances and the lower-triangle representation used by function `dist`[18]

```
x <- matrix(rnorm(30,3), nrow=10)   # so 10 x 3
d <- dist(x)                        # a vector of length 45
dist2sym <- function(d)
{
  n <- attr(d, "Size")
  m <- matrix(0, n, n)
  m[lower.tri(m)] <- d
  m + t(m)
}
sym2dist <- function(m)
  structure(m[col(m) < row(m)], Size = nrow(m))
```

## Dropping indices and levels

Note that by default with our previous example `a[2,,]` is not a $1 \times 5 \times 10$ array but a $5 \times 10$ array. Also `a[2,,1]` and `Xad` are vectors and not arrays. In general if any index range reduces to a single value the corresponding element of the dimension vector is removed in the result. This default convention is sometimes helpful and sometimes not. To override it a named argument `drop=F` can be given in the array reference:

```
sua <- a[2,,]          # a    5x10 matrix
sub <- a[2,,, drop=F]  # a 1x5x10 array
```

Note that the drop convention also applies to columns (but not rows) of data frames, so if subscripting leaves just one column, a vector is returned.

However, this convention does not apply to matrix operations. Thus

```
X <- matrix(1:30, 10, 3)
X[,1]
X %*% c(1,3,5)
```

---

[18]in package `mva` in R.

result in a 10-element vector and a $10 \times 1$ matrix, respectively, which appears inconsistent. (One can argue for either convention, as reducing $1 \times n$ matrices to vectors is often undesirable.) The function `drop` forces dropping, so `drop(X %*% c(1,3,5))` returns a vector and `X[,1, drop=F]` returns a matrix.

S programmers have often overlooked these rules, which can result in puzzling or incorrect behaviour when just one observation or variable meets some selection criterion.

When subsets of a factor are selected by an index vector there is also a `drop` argument available to specify whether the levels are to be pruned to just those which occur in the subset. By default no pruning of levels is done. If you select a subset of the rows of a data frame and you wish the factors to be pruned in this way you must do it separately, for each factor in the data frame.

## 2.4 Vectors, matrices and arrays

### Array arithmetic

Arrays may be used in ordinary arithmetic expressions and the result is an array formed by element-by-element operations on the data vector. The `dim` attributes of operands generally need to be the same, and this becomes the dimension vector of the result. So if `A`, `B` and `C` are all arrays of the same dimensions

```
D <- 2*A*B + C + 1
```

makes `D` a similar array with its data vector the result of the evident element-by-element operations. However the precise rule concerning mixed array and vector calculations has to be considered a little more carefully. From experience we have found the following to be a reliable guide, although it is to our knowledge undocumented and hence liable to change.

- The expression is scanned from left to right.
- Any short vector operands are extended by recycling their values until they match the size of any other operands.
- As long as short vectors and arrays *only* are encountered, the arrays must all have the same `dim` attribute or an error results.
- In the S engines any vector operand longer than all arrays present converts the calculation to one in which all operands are coerced to vectors. A diagnostic message is issued if the size of the long vector is not a multiple of the (common) size of all previous arrays. R gives an error if at any stage in the calculation a vector operand is longer than an array operand.
- If array structures are present and no error or coercion to vector has been precipitated, the result is an array structure with the common `dim` attribute of its array operands.

The new S engine has an additional rule: if any value found has length zero, the result has length zero. (Earlier versions would use the re-cycling rule and fill the result with missing values.)

## Elementary matrix operations

We have seen that a matrix is merely a data vector with a `dim` attribute specifying a double index. However, S contains many operators and functions for matrices; for example `t(X)` is the transpose function.

The operator `%*%` is used for matrix multiplication. Vectors which occur in matrix multiplications are if possible promoted either to row or to column vectors, whichever is multiplicatively coherent. (This may be ambiguous, as we shall see.) Note carefully that if A and B are square matrices of the same size, then A * B is the matrix of element-by-element products whereas A `%*%` B is the matrix product. If x is a vector, then

```
x %*% A %*% x
```

is a quadratic form $x^T A x$, where $x$ is the column vector and $^T$ denotes transpose.

Note that x `%*%` x seems to be ambiguous, as it could mean either $x^T x$ or $x x^T$. A more precise definition of `%*%` is that of an inner product rather than a matrix product, so in this case $x^T x$ is the result. (For $x x^T$ use x `%o%` x.)

The function `crossprod` forms "crossproducts", meaning that

```
XT.y <- crossprod(X, y)
```

calculates $X^T y$. This matrix could be calculated as `t(X)` `%*%` y but using `crossprod` is more efficient. If the second argument is omitted it is taken to be the same as the first. Thus `crossprod(X)` calculates the matrix $X^T X$.

An important operation on arrays is the *outer product*. If a and b are two numeric arrays, their outer product is an array whose dimension vector is obtained by concatenating their two dimension vectors (order is important), and whose data vector is obtained by forming all possible products of elements of the data vector of a with those of b. The outer product is formed by the operator `%o%`:

```
ab <- a %o% b
```

or by the function `outer`:

```
ab <- outer(a, b, "*")
ab <- outer(a, b)          # as "*" is the default.
```

The multiplication function may be replaced by an arbitrary function of two variables (or its name as a character string). For example if we wished to evaluate the function

$$f(x, y) = \frac{\cos(y)}{1 + x^2}$$

over a regular grid of values with $x$– and $y$–coordinates defined by the S vectors x and y respectively, we could use

```
f <- function(x, y) cos(y)/(1 + x^2) # define the function
z <- outer(x, y, f)                  # use it.
```

If the function is not required elsewhere we could do the whole operation in one step using an *anonymous* function[19] as the third argument, as in

```
z <- outer(x, y, function(x, y) cos(y)/(1 + x^2))
```

Matrix multiplication is commonly used for finding row and column sums. Recent versions of S-PLUS have fast internal functions rowSums and colSums   ○ (and colMeans and colVars and so on), and all the engines have rowsum, to do row sums by group.

The function diag either creates a diagonal matrix from a vector argument, or extracts as a vector the diagonal of a matrix argument. Used on the assignment side of an expression it allows the diagonal of a matrix to be replaced.[20] For example, to form a covariance matrix in multinomial fitting we could use

```
> p <- dbinom(0:4, size=4, prob=1/3)   # an example prob vector
> CC <- -(p %o% p)
> diag(CC) <- p + diag(CC)
> structure(3^8 * CC, dimnames=list(0:4, 0:4))   # convenience
      0     1     2     3    4
0  1040  -512  -384  -128  -16
1  -512  1568  -768  -256  -32
2  -384  -768  1368  -192  -24
3  -128  -256  -192   584   -8
4   -16   -32   -24    -8   80
```

In addition diag(n) for a positive integer n generates an $n \times n$ identity matrix. This is an exception to the behaviour for vector arguments; diag(x,length(x)) will give a diagonal matrix with diagonal x for a vector of any length, even one.

## Functions operating on matrices

The standard operations of linear algebra are either available as functions or can easily be programmed, making S a flexible matrix manipulation language. See Section 4.3 of MASS for more details. For reference we note that these functions include solve (including inversion), backsolve, chol (Choleski decomposition), eigen, svd, ginverse, qr, qr.Q, qr.R, qr.coef and qr.resid. These and their statistical applications are discussed in, for example, Thisted (1988) and Gentle (1998).

---

[19] an anonymous function is a argument list and body used inline, without being given a name.
[20] This is one of the few places where the recycling rule is disabled: the replacement must be scalar or a vector of the correct length.

## 2.5   Character vector operations

The form of character vectors can be unexpected and should be carefully appreciated. Unlike say C, they are vectors of character strings, not of characters, and most operations are performed separately on each component.

Note that `""` is a legal character string with no characters in it, known as the empty string. This should be contrasted with `character(0)` which is an empty character vector. As vectors, `""` has length 1 and `character(0)` has length 0.

Character vectors may be created by assignment and may be concatenated by the `c` function. They may also be used in logical expressions, such as `"ann"` < `"belinda"`, in which case lexicographic ordering applies using the ASCII collating sequence in the S engines and the current locale in most ports of R. (The same ordering is used for `sort`.)

There are several functions for operating on character vectors. The function `nchar(text)` gives (as a vector) the number of characters in each element of its character vector argument. The function `paste` takes an arbitrary number of arguments, coerces them to strings or character vectors if necessary and joins them together, element by element, as character vectors. For example

```
> paste(c("X","Y"), 1:4)
[1] "X 1" "Y 2" "X 3" "Y 4"
```

Any short arguments are re-cycled in the usual way. By default the joined elements are separated by a blank; this may be changed by using the argument, `sep=string`, often the empty string:

```
> paste(c("X","Y"), 1:4, sep="")
[1] "X1" "Y2" "X3" "Y4"
```

Another argument, `collapse`, allows the result to be concatenated into a single long string. It prescribes another character string to be inserted between the components during concatenation. If it is `NULL`, the default, or `character(0)`, no such global concatenation takes place. For example

```
> paste(c("X","Y"), 1:4, sep="", collapse=" + ")
[1] "X1 + Y2 + X3 + Y4"
```

Substrings of the strings of a character vector may be extracted (element-by-element) using the `substring` function. It has three arguments

```
substring(text, first, last = 1000000)
```

where `text` is the character vector, `first` is a vector of first character positions to be selected and `last` is a vector of character positions for the last character to be selected. If `first` or `last` are shorter vectors than `text` they are re-cycled in the usual way. If `first` is after `last` or either is missing (`NA`) the corresponding element of the result is an empty character vector.

○      In the new S engine there is a replacement function `substring<-`, so `substring` can be used on the left-hand-side of expressions too.

### Searching and matching

The function `grep` searches for patterns in a vector of character strings, and returns the indices of the strings in which a match was found.   On R and S-PLUS UNIX systems it is based on the utility `egrep`. Except on S-PLUS for Windows,[21] ' . ' matches any character (use '\.' to match  . ) and ' .* ' matches zero or more occurrences of any character, that is any character string.

Simpler forms of matching are done by the functions `match`, `pmatch` and `charmatch`.   Each tries to match each element of its first argument against the elements of its second argument.   The function `match` seeks the first exact match (equality) whereas the other two look for partial matches (the search string starts the matched character string) and match each element in the second argument once only.   All return the value of their `nomatch` argument (which defaults to `NA`) if there is no match.   The function `charmatch` returns the index of a unique match, and `0` if there is more than one match.   With argument `duplicates.ok=F` (the default), `pmatch` returns `nomatch` for duplicate matches, whereas with `duplicates.ok=T` it returns the index of the first match.

*Regular expressions* are powerful ways to match character strings and are familiar to users of such tools as `sed`, `grep`, `awk` and `perl`.   They are used in function `regexpr` (in R and S-PLUS 3.4 and later) which matches one regular expression to a character vector.   The functions `regMatch` and `regMatchPos` of the new S engine have a very similar role, but encode the answer somewhat differently. (See page 97 of MASS for examples.)

R has functions `sub` and `gsub` that look for a match to a regular expression   R in each element of a character vector, and if there is a match, replace the first match ( `sub` ) or all matches ( `gsub` ). Effectively these are implementations of the sed commands

```
s/pattern/replacement/
s/pattern/replacement/g
```

for character vectors. We can write a version of `sub` in S less efficiently by

```
sub <- function(pattern, replacement, x)
{
    m <- regexpr(pattern, x)
    good <- m > 0
    lens <- attr(m, "match.length")
    x[good] <- paste(substring(x, 1, m[good]), replacement,
                    substring(x, m[good] + lens))
    x
}
```

## 2.6   Control structures

Control structures are the commands that make decisions or execute loops.

---

[21]in those versions ' * ' is a wildcard character matching zero or more characters and ' ? ' matches precisely one character.

## Conditional execution of statements

Conditional execution uses either the `if` statement or the `switch` function. The `if` statement has the form

> `if` (*condition*) *true.branch* `else` *false.branch*

First the expression `condition` is evaluated. If the result is true (or non-zero) the value of the `if` statement is that of the expression *true.branch*, otherwise that of the expression *false.branch*. The `else` part is optional and omitting it is equivalent to using "`else NULL`". If *condition* has a vector value only the first component is used and a warning is issued. The `if` function can be extended over several lines, and the statements may be compound statements enclosed in braces `{ }`. Note that there are exceptional rules for parsing `if` and `else`, as conditional evaluation can be done interactively and so the parser must be able to decide if each line is syntactically complete. Thus the expression in the next display could not be used at the top level without the braces, but it can be used without them within a function body (or other braced expression).

Two additional logical operators, `&&` and `||`, are useful with `if` statements. Unlike `&` and `|`, which operate componentwise on vectors, these operate on scalar logical expressions. With `&&` the right-hand expression is only evaluated if the left-hand one is true, and with `||` only if it is false. This conditional evaluation property can be used as a safety feature, as in

```
if (is.numeric(x) && min(x) > 0) { sx <- sqrt(x)
} else  stop("x must be numeric and all components positive")
```

The expression `min(x) > 0` is invalid for non-numeric `x`.

The `if` statement should be distinguished from the `ifelse` function, which is its vector counterpart. Its form is

> `ifelse`(*test, true.value, false.value*)

where all arguments are vectors, and the recycling rule applies if any are short. All arguments are evaluated and *test* is coerced to logical if necessary. In those component positions where the value is true the corresponding component of *true.value* is the result and elsewhere it is that of *false.value*. Since `ifelse` operates on vectors, it is fast and should be used if possible.

Note that `ifelse` can generate `NA` warning messages even in cases where the result contains none. In an assignment such as

```
y.logy <- ifelse(y <= 0, 0, y*log(y))
```

all three arguments to `ifelse` are evaluated, so warning messages may be generated if `y` has any negative components even though the final result will have no `NA` components. Two ways of producing the desired result for vectors `y` of counts are

```
y.logy <- y * log(y + (y==0))
y.logy <- y * log(pmax(1, y))   # alternative
```

○    The `switch` function provides a more readable alternative to nested `if` statements. For example if you wished to allow the user a choice of tests for equality
of variances the conditional statements

```
result <- if (test == "Levene")  levene(y, f)
          else
             if (test == "Cochran") cochran(y, f)
             else bartlett(y, f)
```

would allow three possibilities, Levene's, Cochran's or Bartlett's test (assuming
the three functions were available). This sequence of checks can be replaced by
the construction

```
result <-  switch(test,
               Levene = levene(y, f),
               Cochran = cochran(y, f),
               Bartlett = bartlett(y, f))
```

If the first argument to `switch` evaluates to a character string the value of the
expression is that of the matching named argument, or, if none does match exactly,
that of a final unnamed argument, if any. If the actual argument for the matching
string is missing, `switch` uses the next available argument. For example one
could use

```
result <- switch(test,
    Levene =, levene =, "Levene's test" = levene(y, f),
    Cochran =, cochran =, "Cochran's test" = cochran(y, f),
    Bartlett =, bartlett =, "Bartlett's test" =,
    bartlett(y, f))
```

to allow alternative possibilities for specifying the test to be used, still retaining
Bartlett's test as the 'catch all'. Note that non-standard names may be used if
quoted. Abbreviated names are not matched but we can allow them by using the
function `pmatch` (see page 31).

```
result <- switch(
    pmatch(test, c("Levene", "levene", "Cochran", "cochran"),
           nomatch = ""),
    "1" =, "2" = levene(y, f),
    "3" =, "4" = cochran(y, f),
    bartlett(y, f))
```

Note that the result of `pmatch` is coerced to character mode by the `nomatch`
argument. Here we make use of a default item for a `switch` statement.

The first argument to `switch` may also evaluate to a number, which is coerced to an integer. Argument names are then ignored and the appropriate argument among those remaining, if there is one, is selected. There is no default final
argument; if the number is outside the range 1 to `nargs()-1` the result is NULL.
If a selected argument position is present but the argument itself is vacant there
is no 'drop through' convention. (In this case R gives an error but the S engines    R
return a value of mode `"missing"`, which can result in a puzzling error.)

## Loops: the `for`, `while` and `repeat` statements

A `for` loop allows a statement to be iterated as a variable assumes values in a specified sequence. The statement has the form

```
for(variable in sequence) statement
```

where `in` is a keyword, `variable` is the loop variable and `sequence` is the vector of values it assumes as the loop proceeds. This is often of the form `1:10` or `seq(along=x)` but it may be a list, in which case `variable` assumes the value of each component in turn. The `statement` part will often be a grouped statement and hence enclosed within braces, `{ }`.

The `while` and `repeat` loops do not make use of a loop variable. Their forms are

```
while (condition) statement
```

and

```
repeat statement
```

In both cases the commands in the body of the loop are repeated. For a `while` loop the normal exit occurs when `condition` becomes `F`; the `repeat` statement continues indefinitely unless exited by a `break` statement.

The `next` statement within the body of a `for`, `while` or `repeat` loop causes a jump to the beginning of the next iteration. The `break` statement causes an immediate exit from the loop.

○       S-PLUS also has `For` loops, which 'unroll' the loop and (by default) use a separate process to perform the calculations. See page 157 for some examples.


## 2.7   Vectorized calculations

Programmers coming to S from other languages are often slow to take advantage of the power of S to do vectorized calculations, that is calculations that operate on entire vectors rather than on individual components in sequence. This often leads to unnecessary loops. Explicit loops in S should be regarded as potentially expensive in time and memory use and ways of avoiding them should be considered. (See page 153. Note that this will be impossible with genuinely iterative calculations.)

The functions `apply`, `tapply`, `sapply` and `lapply` offer ways around
R   explicit loops. (R uses a loop internally.) MASS has more examples (on its pages 103–110).

## The function `apply` for arrays

The function `apply` allows functions to operate on successive sections of an array. For example, consider the dataset `iris`[22] which is a $50 \times 4 \times 3$ array of four measurements on 50 specimens of each of three species. Suppose we want the means for each variable by species; we can use `apply`.

The arguments of `apply` are

1. The name of the array, `X`.
2. An integer vector, `MARGIN`, giving the indices defining the sections of the array to which the function is to be separately applied. It is helpful to note that if the function applied has a scalar result, the result of `apply` is an array with `dim(X)[MARGIN]` as its dimension vector.
3. The function, or the name of the function, `FUN`, to be applied separately to each section.
4. Any additional arguments needed by the function as it is applied to each section.

Thus we need to use

```
> apply(iris, c(2,3), mean, trim=0.1)
         Setosa Versicolor Virginica
Sepal L. 5.0025     5.9375    6.5725
Sepal W. 3.4150     2.7800    2.9625
Petal L. 1.4600     4.2925    5.5100
Petal W. 0.2375     1.3250    2.0325
```

where we also show how arguments can be passed to the function, in this case to give a trimmed mean. If we want the overall means we can use

```
> apply(iris, 2, mean)
 Sepal L. Sepal W. Petal L. Petal W.
   5.8433   3.0573    3.758   1.1993
```

Note how dimensions have been dropped to give a vector. If the result of `FUN` is itself a vector of length `d`, say, then the result of `apply` is an array with dimension vector `c(d, dim(X)[MARGIN])`, with single-element dimensions dropped.

## Functions operating on factors and lists

Factors are used to define groups in vectors of the same length as the factor. Each level of the factor defines a group, possibly empty.

In this section we will use the `quine` data frame from the MASS library giving data from a survey of school children. The study giving rise to the data set is described more fully in MASS on page 180; in brief the data frame has four factors, `Sex`, `Eth` (ethnicity—two levels), `Age` (four levels) and `Lrn` (Learner group—two levels) and a quantitative response, `Days`, the number of days the child was away from school in a year.

---

[22] `iris3` in R.

*The function* `table`

It is often necessary to tabulate factors to find frequencies. The main function for this purpose is `table` which returns a frequency cross-tabulation as an array. For example:

```
> attach(quine)
> table(Age)
 F0 F1 F2 F3
 27 46 40 33
> table(Sex, Age)
   F0 F1 F2 F3
F 10 32 19 19
M 17 14 21 14
```

Note that the factor levels become the appropriate `dimnames` attribute for the R frequency array (and in R the names of the factors become the names of the `dimnames` and are printed with the table). If the arguments given to `table` are not factors they are effectively coerced to factors.

*Ragged arrays and the function* `tapply`

The combination of a vector and a labelling factor or factors is an example of what we call a *ragged array*, since the group sizes can be irregular. (When the group sizes are all equal the indexing may be done implicitly and more efficiently using arrays, as was done with the `iris` data.)

To calculate the average number of days absent for each age group we can use the ragged array function `tapply`.

```
> tapply(Days, Age, mean)
      F0     F1     F2     F3
  14.852 11.152 21.05 19.606
```

The first argument is the vector for which functions on the groups are required, the second argument, `INDICES`, is the factor defining the groups and the third argument, `FUN`, is the function to be evaluated on each group. If the function requires more arguments they may be included as additional arguments to the function call, as in

```
> tapply(Days, Age, mean, trim = 0.1)
      F0     F1     F2     F3
  12.565 9.0789 18.406 18.37
```

for 10% trimmed means.

If the second argument is a list of factors the function is applied to each group of the cross-classification. Thus to find the average days absent for age by gender we could use

```
> tapply(Days, list(Sex, Age), mean)
       F0     F1     F2     F3
F 18.700 12.969 18.421 14.000
M 12.588  7.000 23.429 27.214
```

As with `table`, coercion to factor takes place where necessary.

*Operating along structures: the functions* `lapply`, `sapply` *and* `split`

The functions `lapply` and `sapply` are functions in the same family as `tapply` and `apply` for operations on the individual components of lists or vectors. The two functions are called with exactly the same arguments but whereas `sapply` will simplify the result to a vector or an array if possible, `lapply` will always return a list.

The function `split` takes as arguments a data vector and a vector defining groups within it. The value is a list of vectors, one component for each group.

For ragged arrays defined by a single factor the calls `tapply(vec, fac, fun)` and `sapply(split(vec, fac), fun)` usually give the same result. (The only time they will differ is when `sapply` simplifies a list to an array and `tapply` leaves it as a list.) Even though `tapply` actually uses `lapply` with `split` for the critical computations, calling them directly can sometimes be much faster. For example consider the problem of finding the *convolution* of two numeric vectors. If they have components $a_i$ and $b_j$ respectively, their convolution has components $c_k = \sum_{i+j=k} a_i b_j$ . This is the same problem as finding the coefficient vector for the product of two polynomials. A simple function to do this is

```
conv1 <- function(a, b) {
  ab <- outer(a, b)
  unlist(lapply(split(ab, row(ab) + col(ab)), sum))
}
```

Replacing the last line with `tapply(ab, row(ab)+col(ab), sum)` is somewhat slower. We will consider very much faster ways to do this in Chapter 6 using compiled code.

Notice that with `sapply`, `lapply` and `split` if the relevant argument is not a factor or list of factors, once again the appropriate coercions to factor take place.

The first argument to `lapply` or `sapply` may be any vector, in which case the function is applied to each component. For example one way to discover which single-letter names are currently visible on the search path is

```
> Letters <- c(LETTERS, letters)
> Letters[sapply(Letters, function(xx) exists(xx))]
[1] "C" "D" "F" "I" "T" "c" "q" "s" "t"
```

Care is needed when using internal functions[23] such as `exists` and `dim` by name as an argument to `lapply` or `sapply` as the results are not always correct (MASS page 108); using an anonymous function as a wrapper as here *is* safe.

In the new S engine there is a new member of the `apply` family, `rapply`, which is like `lapply` but applied recursively, that is to all lists contained within the supplied list, and to lists within those lists and so on.

-----

[23] and primitive functions in R.

## Frequency tables as data frames

We may find which of the variables in `quine` are factors by

```
> sapply(quine, is.factor)
Eth Sex Age Lrn Days
  T   T   T   T    F
```

since a data frame is a list. Consider the problem of taking a set of $n$ factors and constructing the complete $n$-way frequency table as a data frame, that is, as a frequency vector and a set of $n$ classifying factors. We will work in a way that generalizes to any number of factors.

First we remove any non-factor components from the data frame.

```
quineF0 <- quine[sapply(quine, is.factor)]
```

R     Function `table` in R allows a list as its first argument, but for the other engines (and for other such functions) we need to construct a call with a variable number of arguments. The function `do.call` takes two arguments: the name of a function (as a character string) and a list. The result is a call to that function with the list supplying the arguments. List names become argument names. Hence we may find the frequency table using

```
tab <- do.call("table", quineF0)
```

The result is a multi-way array of frequencies.

Next we find the classifying factors, which correspond to the indices of this array. A convenient way is to use `expand.grid`. This function takes any number of vectors and generates a data frame consisting of all possible combinations of values, in the correct order to match the elements of a multi-way array with the lengths of the vectors as the index sizes. Argument names become component names in the data frame. Alternatively `expand.grid` may take a single list whose components are used as individual vector arguments. Hence to find the index vectors for our data frame we may use

```
> QuineF <- expand.grid(lapply(quineF0, levels))
> QuineF$Freq <- as.vector(do.call("table", quineF0))
> QuineF
  Eth Sex Age Lrn Freq
1   A   F  F0  AL    4
2   N   F  F0  AL    4
3   A   M  F0  AL    5
  ....
```

We use `as.vector` to remove all attributes since some might conflict with the data frame properties.

# Chapter 3

# The S Language:
# Advanced Aspects

In this chapter we consider the details which are not normally important in interactive use of S, and some more formal aspects of the language.

## 3.1 Functions

We have considered functions several times informally in Chapter 2, but we now need to specify more formally how functions can be written and used. Functions in S are created by assignment using the keyword `function`:

```
fname <- function(arg1, arg2, etc.) function.body
```

where `arg1`, `arg2`, `etc`[1] are arguments to be satisfied on the call. The statement, `function.body`, defining the body of the function can be any S statement, but is usually a grouped statement and so enclosed within braces, `{ }`.

Functions can be defined within the body of other functions. This is an important device, both to manage scope issues (see Section 3.4) and to manage the name space. If you define function `myone` in the body of `myfn` then

(i) Calls to `myone` inside `myfn` will use the nested function (unless the programmer is writing really esoteric S code) and

(ii) The function `myone` is not visible outside `myfn` and so will not be accidentally used elsewhere.

This is an under-used feature that C and FORTRAN programmers will benefit from adopting. (We know, it took us a long time to appreciate it.)

A function is called by giving its name with an argument sequence in parentheses

---

[1]Note that we are using `etc.` to denote an indefinite number of similar items, as " ... " is an allowable literal argument with a special meaning. We always use *four* dots when indicating an omission.

39

```
fname(val1, y=val2, etc.)
```

Several protocols are available for specifying both the arguments in the function definition and their values on a call: see the next subsection.

The value returned by the function is the value of the function body, which is usually an unassigned final expression. Alternatively the evaluation of a function may be terminated at any stage by calling the `return` function whose argument specifies the value to be returned. For example if some argument `x` is empty it might be appropriate to return the missing value marker immediately.

```
    . . . .
if(length(x) == 0) return(NA)
    . . . .
```

If `return` is given several, possibly named, arguments the value returned is a list of components with names as supplied,[2] which seems to be an undocumented feature.

## Calling conventions for functions

Functions may have their arguments *specified* or *unspecified* when the function is defined. When the arguments are unspecified there may be an arbitrary number of them. They are shown as ... when the function is defined or printed. Examples of functions with unspecified arguments include the concatenation function, `c(..., recursive=F)`, and the parallel maximum function `pmax(..., na.rm = F)`.

We need to distinguish between the *formal* arguments, those used in the function definition, and the *actual* arguments, those used in the function call.

The complete rules by which formal and actual arguments are matched are

1. First, any actual arguments specified in the `name=value` form where the name *exactly* matches the name of a formal argument are matched. (The supplied `name` is sometimes called the 'tag' in R.)

   In the case of formal arguments that occur *after* a ... argument (as in each of our examples), this is the only way they can be matched.

2. Next, arguments specified in the `name=value` form for which the `name` is a unique partial match for a formal argument name are matched, provided the formal argument does not occur after a ... argument.

3. Any unnamed actual arguments are then matched to the remaining formal arguments one by one in sequence. (This is sometimes called *positional matching*.)

4. All remaining unmatched actual arguments become part of the ... formal argument, if there is one, or else an error occurs.

5. Having unmatched formal arguments is not an error.

---

[2] or the object name if a name is not supplied for that component.

A call to a function may have some initial arguments in positional form but give some later arguments in the named form. For example the two calls

```
t.test(x1, y1, var.equal = F, conf.level = 0.99)
t.test(conf.level = 0.99, var.equal = F, x1, y1)
```

are equivalent.

Functions with specified arguments also have the option of specifying *default values* for those arguments, which are used if a value is not supplied when the function is called. For example, the function t.test [3] has an argument list defined as

```
t.test <- function(x, y = NULL, alternative = "two.sided",
        mu = 0, paired = F, var.equal = T, conf.level = 0.95)
```

so that our previous calls are also equivalent to

```
t.test(x1, y1, , , , F, 0.99)
```

and in all cases the default values for alternative, mu and paired are used. Using the positional form and omitting values, as in this last example, is rather vulnerable to errors and to changes in argument sequence, so the named form is preferred except for the first couple of arguments.

The argument names and any default values for an S function can be found from the on-line help, by printing the function itself or succinctly using the args function. For example in **S-PLUS 3.x**

```
> args(hist)
function(x, nclass, breaks, plot = TRUE, probability = FALSE,
        include.lowest = T, ..., xlab = deparse(substitute(x)))
NULL
```

shows the arguments, their order and those default values which are specified for hist function. (The return value from args always ends with NULL.) Note that even when no default value is specified the argument itself may not need to be specified. If no value is given for nclass or breaks when the hist function is called, default values are calculated within the function.

## The special argument "..."

The 'three dots' argument, " ... ", is special in that any number of arguments may be matched to it on the call. Normally it is simply passed on as an argument to some other function, in which case all corresponding arguments are passed on as they were specified in the original call. (In the hist function they are passed on to barplot.) However, it is also the way to provide a function which can handle a variable number of arguments: see page 46 for ways to write such functions.

Generic functions (see Chapters 4 and 5) make extensive use of " ... " as a way to allow different method functions (versions of the function) to have different sets of arguments.

---

[3] in package ctest in R, with a different default for var.equal.

## Environments

R   User-written R functions have an extra twist, which comes from the use of *lexical scoping* (page 63); they know where they were defined and have access to the R objects available there. It is this feature that necessitates storing collections of R functions together. Its other implications are discussed as they arise.

Technically such R functions are known as *closures*, and they are sometimes referred to as such in the R documentation.

## Function structure

This is a rather more advanced topic, but enables us to explain why the `args` function always prints `NULL`.

Internally, a function is described as a list of arguments and defaults, plus the function's body (an S expression) and in R an enclosing environment. The details of the internal representation differ by engine, but in all of them `args` works by setting the body to `NULL` and then printing the function. (See page 67 for more information.)

In the S engines the arguments plus the body are a single S list and can be ma-
R   nipulated as such. In R there are functions `formals`, `body` and `environment` (and corresponding replacement functions) to manipulate the parts of the representation, and `as.list` and `as.function` to convert functions to and from lists. In the new S engine there are functions `functionArgs`, `functionArgNames` and `functionBody` which can be used to write clearer code, but these are not actually needed.

## 3.2   Writing functions

As a simple example consider a function to perform a two-sample $t$-test.

```
ttest <- function(y1, y2, test = "two-sided", alpha = 0.05)
{
  n1 <- length(y1); n2 <- length(y2)
  ndf <- n1 + n2 - 2
  s2 <- ((n1 - 1) * var(y1) + (n2 - 1) * var(y2))/ndf
  tstat <- (mean(y1) - mean(y2))/sqrt(s2 * (1/n1 + 1/n2))
  tail.area <- switch(test,
    "two-sided" = 2 * (1 - pt(abs(tstat), ndf)),
    lower = pt(tstat, ndf),
    upper = 1 - pt(tstat, ndf),
    {
      warning("test must be 'two-sided', 'lower' or 'upper'")
      NULL
    }
  )
  list(tstat = tstat, df = ndf,
```

```
           reject = if(!is.null(tail.area)) tail.area < alpha,
           tail.area = tail.area)
 }
```

This function requires two arguments to be specified, y1 and y2, the two sample
vectors. It also allows two more, test and alpha, but if no values for these are
given on the call, the default values are used. The result is a list of components
giving information about the test result. This is conventional, although giving
a result as a primary object with additional information carried as attributes is
sometimes appropriate.

It is easy to test the function using simulated data.

```
> x1 <- round(rnorm(10), 1); x1
>  [1]  0.5 -1.4  1.5  0.5 -0.9 -0.9 -0.1 -0.5 -1.2  0.1
> x2 <- round(rnorm(10) + 1, 1); x2
>  [1]  0.4 -0.3  0.9  2.2  1.8  0.7  2.7  1.0  1.1  2.9
> ttest(x1, x2)
$tstat:
[1] -3.6303
$df:
[1] 18
$reject:
[1] T
$tail.area:
[1] 0.0019138
```

A more compact way to print the result is to coerce it to a numeric vector using
unlist:

```
> unlist(ttest(x1, x2))
    tstat df reject tail.area
  -3.6303 18      1 0.0019138
```

## The functions warning and stop

The functions warning and stop are used inside functions to handle unexpected
situations; warning arranges for a warning to be issued when control returns to
the session level but the action of the function continues. For example if we call
our ttest function with an invalid character string for test a warning message
is issued and the default test performed. We use unlist again for a compact
display:

```
> unlist(ttest(x1, x2, test="left"))
    tstat df
  -3.6303 18
Warning messages:
  test must be 'two-sided', 'lower' or 'upper' in: \
      switch(test, ....
```

The function `stop` terminates the action of the function, issues an error message and returns control to the session level immediately. It does not terminate the session as `q()` does. As we shall see in Section 8.2 it can also be made to precipitate a dump of information on the state of the calculation that can often help in tracing errors.

Users can change the behaviour of warnings and errors from the defaults described here: see Section 8.2.

A common idiom is to use `stop("some message")` as a default value when some argument for a function must be supplied. For example the function `rpois` for generating artificial Poisson samples requires that the mean parameter be specified:

```
> args(rpois)
function(n, lambda = stop("no lambda arg"))
NULL
```

## Lazy evaluation

When an S function is called the argument expressions are parsed but not evaluated. When a formal argument is required in a function body the actual argument expression (or the default if it is not supplied) is then evaluated and its value is used by the function. In particular if an argument is not used by the function on a particular call, it is never evaluated and it could involve variables that do not even exist. It merely must parse correctly. This is in striking contrast to most compiled languages where the actual arguments are evaluated before the call to the function is made.

It is important to note that *actual* argument expressions are evaluated in the context of the calling function, known as the *parent frame*, whereas *default* values, if needed, are evaluated in the context of the function itself, known as the *local frame*.

This protocol is called *lazy evaluation* and it has some important consequences for S programming, including

1. Default expressions for formal arguments may involve not only other arguments to the function but also variables local to the function itself. The rule is that the expression must be capable of evaluation when it is needed; in particular any local variables involved must have values at the time of evaluation. Thus it is possible to use a call such as

   ```
   glm.obj <- glm(y ~ x, binomial(link = probit))
   ```

   even though there is no object with name `probit` visible on the search path.

2. Within a function it is possible to retrieve the unevaluated argument expressions using the function `substitute`. The function `deparse` can be used to coerce such an object to mode `character`, which is

essentially the reverse of parsing.   Hence if x is a formal argument
deparse(substitute(x)) gives whatever expression was used for x
and returns it as a character vector (split into several strings if the expression is long).

However, this is not the whole story, and the S and R implementations     R
differ. *When* was the expression used for x? In S this is the actual argument
expression matched to the argument x. In R it is the expression for x
within the local frame. This will give the same answer if x is not altered
within the function body before the call to substitute. In R we may
need to force evaluation of arguments involving substitute, as in the
next example.

The second possibility is often used to generate labels on graphs. For example

```
myplot <- function(x, y, lab = deparse(substitute(y))) {
    lab         # to force evaluation in R
    y <- poly(y, 5)
    ....
    title(main = paste("A plot of", lab))
    ....
}
```

We should note one quirk of lazy evaluation: if the call is to a *generic* function
(as described in Chapters 4 and 5) enough arguments have to be evaluated to
choose a method so evaluation cannot be deferred. This normally applies only to
the first argument.

## Matching argument values and processing "..."

We have already discussed the partial matching of argument *names*. It is also
often convenient with character arguments (such as test in our function ttest)
to provide a set of allowable *values* from which the user may select by partial
matching. The function match.arg(arg, choices) uses the character string
arg to select that value from the character string vector choices for which a
unique partial match occurs. No match or more than one partial match results in
an error.

If the call to match.arg is made from within a function and arg is an argument then choices may be omitted. The default value for arg must then be a
character string vector and this default value is used as the choices argument.
If the function is called and no value for arg is supplied, the first entry in the
default character string vector is the value used. This we could re-write ttest as

```
ttest <- function(y1, y2,
    test = c("two-sided", "lower", "upper"), alpha = 0.05)
{
    test <- match.arg(test)
    ....
```

The function `missing` may be used within a function to check if a value was specified for a particular formal argument. It is often tempting to use `missing` to specify default values of arguments, but the standard default mechanism is usually better. One disadvantage with either scheme is that it may not be possible to pass on 'missingness' to another function, so the convention has arisen to give `NULL` as the default value, and that can be passed down.

A related task is to check whether an argument has been supplied, possibly as part of the ... argument. The new S engine has a function `hasArg` that can be used to check if there is an argument with name given by the argument to `hasArg`. On other systems we need to do this 'by hand'. We can find the current call from within a function by `sys.call`, which gives the actual call, or more usefully by `match.call` which matches the names to the arguments. (In both cases the object is of mode `call`.) Thus within a function we can use

```
Call <- match.call()
hasSep <- pmatch("sep", names(Call), 0) > 0
```

to establish if there was an argument named (exactly or partially matched to) `sep`.

One standard way to process "..." arguments is passing them to `list`. Some recipes for doing so will only work some of the time (for example if the call is not too deeply nested), so it is best to stick to the idea used in the following function, which will find the names of the supplied arguments and attempt to give sensible names to unnamed arguments.

```
nameargs <- function(...)
{
    dots <- as.list(substitute(list(...)))[-1]
    nm <- names(dots)
    fixup <- if (is.null(nm)) seq(along = dots) else nm == ""
    dep <- sapply(dots[fixup], function(x) deparse(x)[1])
    if (is.null(nm)) dep
    else { nm[fixup] <- dep; nm }
}
```

Here `substitute` is playing its usual role: `dots` becomes a list with values the expressions for the ... arguments and names the names (if given) of the arguments. For those arguments without names `deparse` is used to attempt to infer suitable names. Very long results from `deparse` will be split into multiple lines, so we only use the first line.

Function `nargs` returns the number of arguments supplied to the function[4] within whose body it is called. The new S engine also has `nDotArgs(...)`, shorthand for `length(substitute(c(...)))-1`. Note that `c` can be used in place of `list` as here, since we are only interested in giving `substitute` an expression to work on.

---

[4]In the S engines it gives -1 if called outside a function, and 0 in R.

## The `on.exit` function for exit actions

A common use of the " ... " argument is to allow the user to specify temporary changes to the graphical parameters for some plot done within a function. There is a conventional way to do this, namely

1. The " ... " argument is included in the argument list, but only arguments to `par` may be substituted for it on the call.

2. Before any plotting is done the following statements are executed:

   ```
   oldpar <- par(...)
   on.exit(par(oldpar))
   ```

   The first statement records the state of the graphics parameters. The second statement using `on.exit` arranges for the previously current settings to be re-set when the action of the function is terminated (whether normally or by an error exit).

The `on.exit` function can be used in a similar way to allow temporary changes to be made to the `options` settings.

The `on.exit` function has a final argument `add` which allows additional actions to be added to those already specified. This is normally good practice.

This is a place where we can make good use of lazy evaluation.[5] For example

```
testit <- function() {
  errno <- 0
  on.exit(cat("error number was ", errno, "\n"))
  errno <- 2
  invisible(NULL)
}
```

will report 2, as the 'on exit' expression is not evaluated until it is needed.

The current `on.exit` expression can be retrieved by calling `sys.on.exit`. Calling `on.exit()` with no arguments cancels any actions previously requested.

## Saving partial results and restarting

One important application for `on.exit` is to save partial results if a (long) S session should be interrupted. A typical construction is (S engines)

```
on.exit(assign(".partial.res", res[1:(iter-1)],
               where=1, immediate=T))
res <- numeric(lots)
for(iter in 1:lots) res[iter] <- myfun(iter)
```

---

[5]This can be defeated in the new S engine by setting argument `evaluate=T`.

For repetitive calculations that involve applying the same steps to different data sets, some of which may cause an error, there is a function called `restart` that allows the user to over-ride error exits and continue with the calculation by re-calling the present function. *This function should only be used directly with extreme caution and by experienced* S *programmers.*

There is a function, `try`, available in S-PLUS 2000, 5.1 and later that allows users to call `restart` in a controlled environment. A call to an expression is made indirectly by giving it as the argument to `try` as in

```
value <- try(risky(x, y, z))
```

If the computation succeeds the usual value is returned but if it fails the error is trapped and a (possibly system-dependent) error value of class `"Error"` is returned. The user can choose to note the error value and continue the computations. Since an error will have been generated prior to the `restart`, the usual error-reporting mechanism will report an error message to the user.

The idea of `try` can be used in other engines. The code is essentially

```
try <- function(expr, first = T)
{
  restart(first)
  if(first) {
    first <- F
    expr
  } else NULL
}
```

which ensures `restart` is called only once, and returns `NULL` if an error occurs. The argument `first` is not intended for use by the user; it provides a convenient way to count the calls, as when a restart occurs the variables in the function are unchanged (and the arguments are not re-evaluated).

## Writing replacement functions

We saw on page 10 that replacement functions were disguised calls to ordinary functions with names ending in `<-`. The call will be created with the value of the right-hand side of the assignment as the last argument. This argument is conventionally called `value`, and must be called that in R. As a simple example consider

```
"cutoff<-" <- function(x, value) { x[x > value] <- Inf; x }
```

which allows the call `cutoff(x) <- 65`.

Do remember to return the whole altered object.

## Recursion

Functions in S are allowed to be recursive, that is they are allowed to call themselves. This idea is both interesting in itself and useful for an understanding of how S organizes its calculations when functions are called, but in practice it should be used cautiously as it can lead to slow and memory-intensive code. In a few cases it can provide a neat and effective solution to an intrinsically recursive problem.

For example consider the problem of generating all possible subsets of size $r$ from a set of elements of size $n$. A recursive algorithm can be based on the following simple observation. Suppose you single out one of the elements, say the first. Then the subsets of size $r$ consist of those that contain the first and those that do not. The first group may be generated by attaching the first object to each subset of size $r-1$ selected from the $n-1$ others, and the second group consists of all subsets of size $r$ from the $n-1$ others.

If the size of the result is not too large an efficient way to do this is to store the subsets as the rows of an array. The following code will do this:

```
subsets <- function(n, r, v = 1:n)
   if(r <= 0) NULL else
   if(r >= n) v[1:n] else
   rbind(cbind(v[1], subsets(n - 1, r - 1, v[-1])),
                     subsets(n - 1, r     , v[-1]))
```

One potential problem with this is that the name `subsets` occurs in the body of the function definition. If we re-assign the function body to another name and use the name `subsets` for something else, or discard it, our function body will cease to work. It would be useful to have a function body that continued to work regardless of the name given it. This is the purpose of the function `Recall`. An amended function which uses an internal function to do the work (and thereby checks the arguments only once) is

```
subsets <- function(n, r, s = 1:n) {
   if(mode(n) != "numeric" || length(n) != 1
      || n < 1 || (n %% 1) != 0) stop("bad value of n")
   if(mode(r) != "numeric" || length(r) != 1
      || r < 1 || (r %% 1) != 0) stop("bad value of r")
   if(!is.atomic(s) || length(s) < n)
     stop("s is either non-atomic or too short")
   fun <- function(n, r, s)
     if(r <= 0) vector(mode(s), 0) else if(r >= n) s[1:n] else
     rbind(cbind(s[1], Recall(n - 1, r - 1, s[-1])),
            Recall(n - 1, r, s[-1]))
   fun(n, r, s)
}
> subsets(5, 3)
        [,1] [,2] [,3]
[1,]    1    2    3
[2,]    1    2    4
```

```
 [3,]   1   2   5
 [4,]   1   3   4
 [5,]   1   3   5
 [6,]   1   4   5
 [7,]   2   3   4
 [8,]   2   3   5
 [9,]   2   4   5
[10,]   3   4   5
```

Of course if the size of the result is too large it may have to be generated sequentially or in smaller groups.

Note that in S (but not in R) if a function is defined within another (as `fun` in our last example) it exists in a frame that is not on its own search path. This implies that the only simple way it can be recursive is to use `Recall`. Note also that `Recall` is not a standard function object and cannot, for example, be used as the function argument for the `apply` family.

### Providing vectorized functions

The S mathematical functions such as `sin`, `log` and `dnorm` have the useful property that if their first argument is a vector, the result is a vector of the same length. In some functions, such as `ifelse` or `pnorm`, the recycling rule applies to all arguments so the length of the result is the length of the longest argument.

```
> pnorm(1, mean = 0, sd = 1:5)
[1] 0.84134 0.69146 0.63056 0.59871 0.57926
```

In designing a general-purpose function some care should be given to making it conform to this convention. (For some S functions if the argument is a matrix or array, so is the result.)

We can illustrate some of the techniques of writing such vectorized functions by the digamma function. This function is usually written as $\psi(z)$ and is defined by

$$\psi(z) = \frac{d}{dz} \log \Gamma(z)$$

It occurs in statistics in several places, such as estimating the parameters of the gamma distribution or fitting the negative binomial distribution. (It is also used in our function `qda`.) We will write a function that accepts real or complex arguments, but unless the real part is positive the result will be `NA`.

If the real part, $\text{Re}(z)$, is large enough, say larger than 5, the result may be accurately calculated by an asymptotic expansion derived from Stirling's formula for the gamma function (Abramowitz & Stegun, 1965), namely

$$\psi(z) = \log z - \frac{1}{2z} - \frac{1}{12z^2} + \frac{1}{120z^4} - \frac{1}{252z^6} + \frac{1}{240z^8} - \frac{1}{132z^{10}}$$
$$+ \frac{691}{32760z^{12}} - \frac{1}{12z^{14}} + \frac{3617}{8160z^{16}} + O(z^{-18})$$

For smaller values of $\mathrm{Re}(z)$ there is a recurrence formula that may be used to relate the calculation to the case of larger $z$:

$$\psi(z) = \psi(z+k) - \frac{1}{z} - \frac{1}{z+1} - \cdots - \frac{1}{z+k-1}$$

Since we only cater for the case $\mathrm{Re}(z) > 0$ we use this formula with $k = 5$ for small $\mathrm{Re}(z)$.

An S function (in library MASS, adopted for S-PLUS 2000) for $\psi(z)$ that uses these ideas is

```
digamma <- function(z) {
  if(any(omit <- Re(z) <= 0)) {
    ps <- z; ps[omit] <- NA
    if(any(!omit)) ps[!omit] <- Recall(z[!omit])
    return(ps)
  }
  if(any(small <- Re(z) < 5)) {
    ps <- z
    x <- z[small]
    ps[small] <- Recall(x + 5) - 1/x - 1/(x + 1) -
                 1/(x + 2) - 1/(x + 3) - 1/(x + 4)
    if(any(!small)) ps[!small] <- Recall(z[!small])
    return(ps)
  }
  x <- (1/z)^2
  tail <- x * (-1/12 + x * (1/120 + x * (-1/252 +
          x * (1/240 + x * (-1/132 + x * (691/32760 +
          x * (-1/12 + 3617 * x/8160)))))))
  log(z) - 1/(2 * z) + tail
}
```

The strategy is

1. Check if any components of $z$ have a non-positive real part and make the result NA in those positions. Fill any other components by a recursive call.

2. Check if any components of $z$ have a small real part. If so fill the corresponding components of the result by a recursive call using the recurrence formula and the other components using a recursive call.

3. Finally implement the truncated asymptotic expansion. Use a nested form to avoid calculating large negative powers and to reduce round-off error through underflow.

Remember that where statements are spread over two lines the first line *must* be syntactically incomplete. For example if in the two lines

```
ps[small] <- Recall(x + 5) - 1/x - 1/(x + 1) -
             1/(x + 2) - 1/(x + 3) - 1/(x + 4)
```

the minus sign from the end of the first line were transferred to the beginning of the second, both expressions would be evaluated separately and not combined, and the result of the function would be erroneous where this branch was used. This is a common kind of error, often very difficult to locate. In extreme cases the entire right-hand side might be enclosed within parentheses for safety's sake. When S reproduces an expression such as the nested product term `tail` of the function above in printed code, unnecessary bracketing is often generated, presumably to insure against this kind of error.

## 3.3   Calling the operating system

Sometimes there is an operating-system command that will do exactly what you want. Functions are provided for communication with operating-system commands in each of the systems. A few are common to all: for example `proc.time` returns a vector giving the times since the beginning of the session. However, even here what the output means differs by system.

### S-PLUS under UNIX

The syntax of the `unix` command is

```
unix(command, input = NULL, output.to.S = T)
```

The argument `command` is a character string which is passed to a standard shell to execute. (A variation on the `unix` function, `unix.shell`, allows the shell to be specified by the argument `shell`. It is reasonable to assume that standard Bourne shell commands will work on all platforms.) If `input` is specified, it is written to a file used for standard input to the command. By default the standard output is returned line-by-line as a character vector, but if the command needs to interact with the user, supply the argument `output.to.S=F` when the exit status of the command is returned (as returned by the `system` call, normally 256 times the usual exit status, so zero indicates success).

   For example, we may test if a file exists and is readable, before attempting to read data from it, using

```
file.exists <- function(name)
    unix(paste("test -r", name), output.to.S = F) == 0
```

   Useful functions in conjunction with `unix` are `tempfile` which creates a unique name for a temporary file (but not the file itself), `dput` and `dget` to write and read objects and `unlink` to remove files. The functions `tempfile` and `unlink` are themselves simple calls to `unix`:

```
tempfile <- function(pattern = "file")
    paste("/tmp/", pattern, unix("echo $$"), sep = "")
unlink <- function(x) # S-PLUS 3.x version
    invisible(unix(paste("rm", paste("'", x, "'", sep = "",
                collapse = " ")), output = F))
```

We use these to write a function to compare two objects (a simpler version of the function `objdiff`):

```
objchk <- function(x, y) {
    old <- tempfile("old"); dput(x, old)
    new <- tempfile("new"); dput(y, new)
    on.exit(unlink(c(old, new)))
    unix(paste("diff", old, new, "| less"), output = F)
    invisible()
}
```

A good use of `unix` is to manipulate character strings; for example, to convert to upper or to lower case we could use the functions

```
to.upper <- function(str) unix('tr "[a-z]" "[A-Z]"', str)
to.lower <- function(str) unix('tr "[A-Z]" "[a-z]"', str)
```

which makes use of the `input` argument.

One very useful function based on operating system calls is `unix.time`, which times its argument and returns a 5-element vector given by the UNIX command `time`: the times (in seconds) of the user, system and elapsed times, together with the user and system times taken by child processes (if any). Further, although this is a function, it is arranged so that assignments within it are done at the level of its call. Note that if the argument returns an invisible result, this results in the return value from `unix.time` being marked as invisible and not printed. (If this happens unintentionally, use `print(.Last.value)`.)

S-PLUS 5.x introduced a `shell` function, which is like `unix` but with a `mustWork` argument for safety purposes, and `pipe` which allows commands to be sent via a pipe to another process. Users should consult the on-line help information for more details.

## S-PLUS under Windows

The Windows interface function is

```
system(command, multi = F, translate = F, minimized = F)
```

and the arguments `multi` determine whether the command is to be run concurrently or waited for, and `translate` whether file paths are converted between UNIX–style `"/test/file"` and MS-DOS–style `"\test\file"`. Note that this calls the command directly: no command interpreter / shell is used.

There is also an MS-DOS interface function

```
dos(command, input, output = T, multi = F, translate = F,
    redirection = F, minimized = F)
```

By default the `input` is appended to the command, but if `redirection = T`, redirection is used (as with the `unix` command under UNIX). Using `multi = T` sets up a MS-DOS box for the command which may need to be closed by the user.

For example we could use `dos` to write a `file.exists` function:

```
file.exists <- function(name)
    length(dos(paste("attrib", name))) > 0
```

but this operation is already available using the S-PLUS system function `access`.

The function `dos.time` is the equivalent of `unix.time`, but measures elapsed time (in seconds) only.

### R

The interface command in all versions of R is `system`, with syntax under UNIX

```
system(call, intern = FALSE, ignore.stderr = FALSE)
```

The argument `intern` behaves like `output.to.S` under S-PLUS. If the argument `ignore.stderr` is true (error) messages written to `stderr` are discarded. The platform's standard shell is used to execute the command.

The Windows version allows a little more control, having arguments

```
system(command, intern = FALSE, wait = TRUE, input,
        show.output.on.console = FALSE,
        invisible = FALSE, minimized = FALSE)
```

Arguments `minimized` and `invisible` apply to the window in which the command is run (if separate). Argument `wait` is equivalent to not-multi, and if `show.output.on.console` is true (and `intern` is false) the output is listed in the console rather than in a separate, transient, window. No command interpreter or shell is used, but there is a function `shell` to execute commands under a shell.

Timing can be performed by `system.time` or, for compatibility with S-PLUS, `unix.time` (on all platforms).

## 3.4   Databases, frames and environments

This section is an area of fundamental differences between the S engines and R. We change our notation here: for this section only 'S' does not include R systems.

### Overview

A *frame* associates names with values, here the names of S objects with their internal representations. S stores objects both in frames in (virtual) memory and in databases which may be files or directories on the disk system. There is an ordered list of frames and databases that are searched for S objects by name, as shown in Figure 3.1. Precisely how those frames and databases get to be on the search list and the order they are searched is the main subject of this section.

R works with *environments*, all of which are in (virtual) memory, although the objects in an environment can be saved to and re-loaded from files. An environment is a frame plus an *enclosure*, a reference to another environment, so

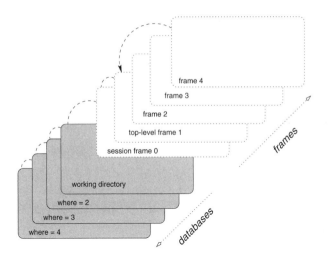

**Figure 3.1**: Illustration of the S search path from frame 4 (the dashed line) with the hierarchy of frames (white) and databases (grey). This figure does not apply to R.

environments are organized as a tree. (Some of the documentation uses 'environment' in a less formal sense as the chain of frames found by starting at an environment and following the references.)

In each engine objects created at the top level end up in a default place, the *working directory* in S and the *workspace* in R. Usually this is all the end user needs to know about, and the contents[6] of the default place are listed by `ls()` or `objects()`.

The only other objects most users will care about are those provided in libraries attached by calls to `library` or always attached at startup. These come after the 'default place' in the search path.

Some functions to access databases are given in Table 3.1. Understanding these is important, as the engine implicitly uses `get` to find objects and `assign` to store them.

Most programmers will need to understand a little more and as functions such as `library` are themselves written in S and not always very well documented, some programmers will need to know a *lot* more. All programmers need some conception of where objects will be found, to avoid surprises when the wrong one is found or when no match is found at all.

We know this section is hard work, but it is essential to achieve a mastery of S programming.

### The search path

We start by considering the databases, the grey objects in Figure 3.1, and how objects are found from the top level (the prompt in an interactive session).

---

[6]not quite: by default R follows UNIX and omits from the list objects whose name starts with a dot.

*Search paths in* S

Objects stored permanently in S are kept in one of the databases on disk, either
as separate files within a directory (a .Data directory or perhaps _Data under
Windows) or as part of a large file named as __BIG or __Objects within such a
directory.

When S looks for an object named at the command line, it searches through a
sequence of places known as the *search path*. The names of the places currently
making up the search path are given by invoking the function

```
search()
```

Usually the first entry in the search path is the data subdirectory of the current
working directory.

The names of the objects held in any place in the search path can be displayed
by giving the objects function an argument. For example,

```
objects(2)
```

lists the contents of the database at position 2 of the search path. It is also possible to list selectively, using the pattern and test arguments of objects
but as how to use this differs between systems, please consult your on-line help.
Omitting the argument lists objects in database 1.

Conversely the function find(*object*) discovers where an object appears on
the search path, perhaps more than once. For example (on one of our UNIX systems after library(treefix, first=T) ) we have

```
> search()
[1] ".Data"
[2] "/usr/local/splus/library/treefix/.Data"
[3] "/usr/local/splus/splus/.Functions"
[4] "/usr/local/splus/stat/.Functions"
[5] "/usr/local/splus/s/.Functions"
[6] "/usr/local/splus/s/.Datasets"
[7] "/usr/local/splus/stat/.Datasets"
[8] "/usr/local/splus/splus/.Datasets"
> find("prune.tree")
[1] "/usr/local/splus/library/treefix/.Data"
[2] "/usr/local/splus/splus/.Functions"
```

(The format will differ between systems and S-PLUS versions. In particular, with
the new S engine it will be like

```
> search()
[1] ".Data"           "treefix"         "splus"      "stat"
[5] "data"            "documentation"   "trellis"    "main"
> find("prune.tree")
[1] "treefix" "splus"
```

**Table 3.1**: Functions to access arbitrary databases, frames and environments

| | |
|---|---|
| `assign` | Creates a new `name=value` pair in a specified database. |
| `exists` | Tests whether a given object exists. |
| `get` | Returns a copy of the object, if it exists; otherwise returns an error. |
| `objects` | Returns a character vector of the names of objects in the specified database. |
| `remove` | Deletes specified objects, if they exist, from the specified database. |

although the full path information is available via `search("paths")`.)

In the example above the object named `prune.tree` occurs in two places on the search path. Since our version from the `treefix` library occurs in position 2 before the S-PLUS version in position 3, ours is the one that will be found and used. By this mechanism changes may be made to S-PLUS functions without affecting the original version.

The 'places' on the search path can be of two main types. (There are others.) As well as data directories of (S created) files, they can also be S lists, usually data frames. In the S literature any entity that can be placed on the search path is sometimes referred to as a *database*. The database at position 1 (normally `.Data` or `_Data`) is called the *working directory* or *working database*.

If several different objects with the same name occur on the search path, all but the first will be masked and normally unreachable. To bypass this, the function `get` may be used to select an object from any given position on the search path. For example,

```
s.prune.tree <- get("prune.tree", where = 3)
```

copies the object called `prune.tree` from position 3 on the search path to an object called `s.prune.tree` in the working database.[7] Functions `masked` and `conflicts` are available to help check if system objects are being masked by user-defined objects, whether intentionally or not.

The function `exists` checks if an object exists on the search path, or in a specific database if argument `where` is given. Both `get` and `exists` have a `mode` argument that can be used to restrict the search to objects of a specific mode, and the new S engine has the specialization `getFunction` to look for functions, with more options (for example to ignore generic functions).

Extra directories, lists or data frames can be added to this list with the `attach` function and removed with the `detach` function. Examples were included in the introductory session. Normally a new entity is attached at position 2, and `detach()` removes the entity at position 2, normally the result of the last `attach`. All the higher-numbered databases are moved up or down accordingly. Note that lists can be attached by `attach(alist)` but, if detached by name, must[8] be detached by `detach("alist")`. If a list is attached, a copy is used,

---

[7]Note that the object name as the argument to `get` must be given in quotes.

[8]in the old S engine. In the new S engine `detach(alist)` *is* allowed.

so any subsequent changes to the original list will not be reflected in the attached copy. When a list is detached the copy is normally discarded, but if any changes have been made to that database it will be saved unless the argument `save=F` is set. The name used for the saved list is of the form `.Save.alist.2` (and is reported by `detach`) unless argument `save` is a character string when the database will be saved under that name.

When a command would alter an object which is not on the working database, a copy must be made on the working database first. The old S engine did this silently, but the new S engine does not and will report an error, so a manual copy must be made. Objects are usually altered through assignment with a replacement function, for example (MASS, page 165)

```
> hills <- hills
> hills$ispeed <- hills$time/hills$dist
```

A frequently asked question is how to assign to an object whose name has been computed and is held in a character string. This is one purpose of the function `assign`, for example

```
for(power in 1:5) assign(paste("x", power, sep=""), x^power)
```

Function `assign` can also be used with its `where` argument to put (or replace) objects in other databases, although for system databases this needs file system permissions that most users will not have. There is an argument `immediate` which defaults to false, but if set to true forces the assignment immediately, circumventing the backout mechanism that S uses for safety. This is useful in error-handling code (as used on page 47) and sometimes to save memory.

To remove objects permanently from the working database the function `rm` is used with arguments the names of the objects to be discarded, as in

```
rm(x, y, z, ink, junk)
```

If the names of objects to be removed are held in a character vector it may be specified as a named argument. An equivalent form of the preceding command is

```
junklist <- c("x", "y", "z", "ink", "junk")
rm(list = junklist)
```

The function `remove` can be used to remove objects with non-standard names or from data directories other than position 1 of the search path. The objects to be removed must be specified as a character vector. A further equivalent to the preceding command is

```
remove(junklist, where = 1)
```

To remove all the objects in the working directory use `remove(objects())`.

A considerable degree of *caching* of databases is performed, so their view may differ inside and outside the S session. To avoid this, use the `synchronize` function. With no argument, it writes out objects to the current working database. With a numerical vector argument, it re-reads the specified directories on the search path, which can be necessary if some other process has altered the database.

*Search paths in* R

The search path in R is also given by the `search` command, for example

```
> search()
[1] ".GlobalEnv"    "Autoloads"     "package:base"
```

These are environments stored in virtual memory. The first is the user's workspace and the last is where the functions that implement R (and are written in R) are stored.

At the top level objects are searched for in each environment in turn, and (by default) objects created at the top level are stored in the workspace.

The function `attach` can be used to add 'databases' to this list and they can be removed with the `detach` function. In practice this limits the possibilities to either list-like R objects such as data frames or new environments attached indirectly via the `library` function (which creates an environment and effectively sources a file of R code to create objects in that environment). By default the databases are attached at position 2, but this can be changed by the `pos` (not `where`) argument. Unlike S, databases cannot be attached at position 1. Thus the 'working database' is always the session workspace, but objects can be written into other databases by using `assign` with a `pos` (not `where`) argument. (If the database is a list-like object, the original object is unchanged.) Databases can be detached by position or by name, and the name given to `detach` does not need to be quoted (but can be).

Functions `get` and `exists` work in the same way as S (including using `where`).

The functions `rm` and `remove` are the same function in R; they behave like `rm` in S but also have a `pos` argument. To remove all objects in the workspace use `rm(list=ls(all=TRUE))`.

As R databases are held in memory, there are no caching and synchronization issues. Objects must be written out to disk by (implicit or explicit) calls to `save`.

## The evaluation model in S

The details of searching are slightly different when S is already evaluating a function or expression, and the frames shown in white on Figure 3.1 must also be considered. An *evaluation frame* is a list that associates names with values. Their purpose is much the same as databases on the search path. The frames are numbered, and one has to be careful not to confuse the numbering of frames and of the search list.

In most cases[9] when S sets out to evaluate an expression or function call, it generates a new frame for that evaluation, called the *local frame*. (It gets the first available frame number, but frames are not usually referred to by number. The top-level expression gets frame 1.)

---

[9]the exceptions are simple internal calls, where 'simple' is not documented.

When an S expression is evaluated within a frame, any values needed are obtained by reference to that frame *first*. To show this in action we can use the function `eval` which takes an S object of mode `expression` and (optionally) a local frame:

```
> a <- 1;  b <- 2;  d <- 3
> eval(expression(a + b + d), local = list(a=10, b=20))
[1] 33
> a + b + d
[1] 6
```

Any names (like d in this example) not satisfied by the local frame are next referred to the frame of the top-level expression, frame 1, then the frame of the session, frame 0, and then on to the search path. The sequence of databases consisting of the local frame, frame 1, frame 0 and the search path may be called the *search path* for an object. Notice that the search path for an object at, say, frame 6, does *not* include frames 2 to 5; this can cause unexpected problems if not borne in mind. In particular, objects created in parent functions will *not* be found in this search. This aspect causes puzzlement so frequently that we emphasize it:

> The S scope rules
> *Objects not found in the current frame are next referred to the frame of the top-level expression, frame 1, then the frame of the session, frame 0 and then on to the search path. Objects created in parent functions* will not *be found in this search.*

See Figure 3.1 (page 55) for a pictorial representation.

A small example where it sometimes seems most puzzling may underline the point. Consider the following elementary function that performs a regression using a simplified[10] Box and Cox transformation, defined by $y^\lambda$ if $\lambda \neq 0$ or $\log y$ if $\lambda = 0$.

```
bcreg <-  function(x, y, lambda)
{                                   # WARNING: errors present
    bc <- function(y)
        if (abs(lambda) < 0.001) y^lambda else log(y)
    lsfit(x, bc(y))
}
```

This fails because within the locally defined function `bc` the argument `lambda` of the enclosing function is not visible. There are two ways to correct it. The preferred way is to include `lambda` as a second argument to the internal function `bc` and on the call, but a second way is to get it directly:

```
bcreg <-  function(x, y, lambda)
{
    bc <- function(y) {    # WARNING: artificial
        lambda <- get("lambda", frame = sys.parent())
```

---

[10]for the full version see MASS, page 182.

```
        if (abs(lambda) < 0.001) y^lambda else log(y)
    }
    lsfit(x, bc(y))
}
```

This is definitely artificial and not recommended, but it does illustrate that with low-level tools like `get` variables may be obtained from any frame or position on the search path.

Frame 0 is born with the session and dies with the session. New entries may be added to this frame, but unlike ordinary assignments made during the session they are not permanent. To be permanent an object must reside in an object database on the search path. Frame 0 is used by the system, for example for storing options set by `options` in object `.Options`.

When an expression is evaluated at the top level it (normally) initiates a new frame called frame 1, usually containing only the unevaluated expression and an auto-print flag specifying whether the result is to be printed or not.

```
> objects(frame=0)
[1] ".Device"            ".Devices"            ".Options"
[4] ".PostScript.Options"
> objects(frame=1)
[1] ".Auto.print" ".Last.expr"
```

Frame 1 is special. By the rules we have given so far

```
{ x <- 3; y <- 5; 7 }
```

is an expression which has value 7 and will be evaluated in frame 1. Thus the assignments of `x` and `y` take place in frame 1, and not the working database. However, the objects `x` and `y` do appear in the working database, but only if the expression is executed without any errors. We can find the number of the current frame by calling `sys.nframe` and see what is in the current frame from `sys.frame`

```
{ x <- 3; y <- 5; print(sys.nframe()); print(sys.frame()); 7 }
```

Try it! The answer depends on the engine: see Becker *et al.* (1988, p. 347) and Chambers (1998, p. 178).

If the expression that generated frame 1 contains function calls, these generate frames at higher levels as they are processed. These frames contain the names and values of arguments and local variables within the function. Normal assignments within functions only change the local frame, and so do not affect values for variables outside that frame. This is how recursion is accommodated. Each recursive call establishes its own frame without altering values for the same variables at different recursion levels. Note that a function call at the top level generates a frame 1 for the expression and then a frame 2 for the body of the function. Frames generated by function calls initially contain the values of the actual arguments (but whether they contain explicitly arguments with default values or ... depends on the engine).

The functions in Table 3.1 (page 57) can be applied to frames as well as databases, by specifying a particular frame number as argument `frame`. The functions `get` and `exists` by default consult the entire search path starting with the local frame. The others consult the local frame (or the working database at the top level) unless another focus is specified. Functions `get` and `exists` can be used with argument `inherit = T` when they provide exceptions to the scope rules and do search all the parent frames as well as the databases on the search path.

There is a special assignment operator, `<<-`, which can be used within a function to make an assignment to the working database. If the user of the function is unaware that such an assignment is going to be made it can have disastrous results. We discourage use of this 'superassignment' operator as a general practice, and particularly for functions to be made publicly available, except when the primary purpose of the function is to achieve the assignment side effect such as `source` and `fix`.

As another example, the `get` function can be made to 'look at itself':

```
> get(".Last.expr", frame=1)
expression(get(".Last.expr", frame = 1))
> get(".Auto.print",frame=1)
[1] T
```

Notice that the object `.Last.expr` in frame 1 can be used to discover the interactive level expression that enacted the present expression. This can sometimes be useful, for example, if it is necessary to decide if the present top-level expression is an assignment or not.

In frame 1 automatic printing is normally turned on. If an expression or the return value of a function is given as `invisible(value)` it is not automatically printed. The `invisible` function itself is quite short:

```
> invisible
function(x = NULL)
{
   assign(".Auto.print", F, frame = sys.parent(2))
   x
}
```

The function `sys.parent` takes an integer argument n and returns the number of the frame n generations behind the present. The assignment inside `invisible` simply turns off auto-printing in the frame of the parent of the function or expression that called `invisible`. Other functions whose names begin with "`sys.`" are available to determine other aspects of the current memory frames. See the on-line help for `sys.parent` for more details.

*Using frame 1*

Sometimes it is necessary for variables in a function to be made available to functions it calls. The easiest way to do this is to assign the variables in frame 1, since this is part of the globally visible search path throughout the evaluation of the current top-level expression. Care must be taken not to mask other variables further along the search path unintentionally. One way of minimizing the chance of this is to use some naming convention for such temporary variables so that they are unlikely to conflict. Starting the name with a period is a common convention.

## Lexical scoping and the R evaluation model

The ways R finds objects are similar to S, and functions with names such as get and assign are available, but this similarity hides fundamentally different strategies.

When asked to evaluate a call to a function (but *not* a braced expression), R creates a new environment. All the functions[11] in Table 3.1 can be applied to environments generated during evaluation as well as places on the search path, by using the argument envir (which should be an environment object, not the number of an environment). Apart from objects, they have an argument inherits (note this is plural, unlike S) which defaults to false for assign and remove and true for exists and get. When inherits is true, the name is searched for in the specified environment and then its enclosures up to the workspace then along the search path until the name is matched (or not). In the case of assign with inherits = TRUE, if no match is found assignment takes place in the workspace.

R has a <<- assignment operator whose semantics differ from those of S. It is equivalent to calling assign with inherits = TRUE and envir the enclosing environment. That is, the replacement occurs in the nearest enclosing environment that contains an object of the same name, or the workspace if none do.

R handles expressions such as

```
{ x <- 3; y <- 5; print(ls()); does.not.exist }
```

rather differently from S: it does not use a separate evaluation environment, and so does assign x and y even though evaluation of the expression fails. (Try printing sys.nframe() : it reports 0 .)

*Who is the parent anyway?*

'Lexical scoping' comes into play when we consider how R finds objects to associate with names when evaluating a function. In principle the answer is easy: it searches the 'environment', that is in the frame of the environment, then in the frame of the enclosing environment, then the frame of the enclosing environment of the enclosing environment, and so on. In less precise language, it searches up the tree of environments until it finds a match or reaches the root, which will be the user's workspace, and then along the search path.

---

[11] and also ls, unlike S.

So, we need to know where the evaluation environment is on the tree of environments. When the evaluation environment is a function body, the enclosing environment is that of the function, which is normally the environment within which the function was defined. This is what is meant by *lexical* scoping.

> The *enclosing* environment need not be the *parent* environment, the environment from which the evaluation environment was created, and is usually the environment within which the function was created, often the workspace.

Notice that the end effect is somewhat similar to that of S; the frames of the parents of a function are not normally searched, so the search goes from the local environment to the user's workspace and then along the search path given by search() (as the enclosing environment of all functions in packages is set to the workspace). However, the scope is different for functions defined within other functions. (A function's environment can be changed by using environment as a replacement function, a complication we will not explore here.)

As a consequence of lexical scoping the first version of bcreg (page 60)

```
bcreg <- function(x, y, lambda)
{                               # works in R but not S
  bc <- function(y)
    if (abs(lambda) < 0.001) y^lambda else log(y)
  lsfit(x, bc(y))
}
```

does work in R, as lambda is in the environment of bc.

R has functions such as sys.parent and sys.parents which return the environment numbers of the parent n generations back or of all parents. Note that these are of the *parent*, not the enclosing environment. Since assign and eval want an environment for their argument envir, not an environment number, there is frequent use for the expression sys.frame(sys.parent(n)) which can be replaced by a call to the convenience function parent.frame(n).

Note too that there is no equivalent of frame 1 which is always visible during a top-level expression: only the workspace (which has environment number 0) is guaranteed to be visible. However, by using locally defined functions we can wrap up variables we need to keep visible with a function, as they will be in the defining environment. There are two occurrences of assignment to frame 1 in our S libraries. That in loglm in library MASS is used to record the original call and formula before a call to UseMethod. Here the wrapping trick is fairly straightforward: we just make the worker function loglm1 and its methods local functions. We *do* need to make the methods local, as method dispatch does not pass on the environment.

```
loglm <-
  function(formula, data = sys.frame(sys.parent()), subset,
           na.action, ...)
  {
```

```
    loglm1 <- function(formula, data, ...)
              UseMethod("loglm1", data)
    loglm1.default <-
      ....

    .call <- match.call()
    if(missing(data) || inherits(data, "data.frame")) {
        m <- match.call(expand = FALSE)
        m$... <- NULL
        m[[1]] <- as.name("model.frame")
        data <- eval(m, sys.frame(sys.parent()))
        .formula <- as.formula(attr(data, "terms"))
    } else {
      trms <- attr(data, "terms") <-
        terms(formula <- denumerate(formula))
      .formula <-  renumerate(as.formula(trms))
    }
    loglm1(formula, data, ...)
  }
```

The other example is function Strauss in library spatial. The idea there
is to have a persistent state across calls within a single top-level expression only.
There seems no easy way to do that in R, and the strategy was somewhat dangerous.

We will see more ways to make good use of R's scope rules on page 95 and
in Chapter 7.

## 3.5   Computing on the language

Everything in S is an object. This includes the expressions that make up the
language itself so it is possible to compute these expressions with the language.
Doing this directly can be rather tricky, but fortunately several tools are available
to assist. It is helpful to know how the expressions are represented, and this is
shown in Table 3.2. Note that a function in an S engine is just a list with a special
structure and mode function; this does not apply in R.

### Computing commands

We often see a need to construct a command to evaluate next depending on the
context.

The simplest idea is to think 'if I were at the command line, I would know
what to type', and to simulate typing zz <- 63 at the command line by

```
varname <- "zz"; value <- 63
cmd <- paste(varname, "<-", value)
eval(parse(text=cmd))
```

**Table 3.2**: S expressions as S objects. The S expression shown is represented as a vector of the form shown and of the mode given. Based on Becker *et al.* (1988, p. 234). Note that operators are function calls and are represented as such.

|            | expression      | vector           | mode       |
|------------|-----------------|------------------|------------|
| function   | `function(a, b)`| `(a, b, body)`   | `function` |
| function call | `fn(x, y)`   | `(fn, x, y)`     | `call`     |
| operator call | `x op y`     | `("op", x, y)`   | `call`     |
| braces (S) | `{x ; y}`       | `(x, y)`         | `{`        |
| (R)        |                 | `("{", x, y)`    | `call`     |
| formula    | `lhs ~ rhs`     | `("~", lhs, rhs)`| `call`     |

This can indeed be useful, but it is not a good way to do this example: a better way is to use `assign("zz", 63)`.

It will be helpful to understand the terms used. Here `parse` takes a character string (which can include newlines or semicolons and so represent several S commands) or character vector and turns it into an object of mode `expression`. Function `eval` then takes the expression and evaluates it in the local frame. The function `deparse` reverses the parsing process and gives a character vector (not necessarily a single string) representing the expression.

Another frequent use of computed expressions is to assemble a function call with a list of arguments that depend on the context. Again, there are often easier ways to do this. One is the function `do.call` that we saw on page 38. This takes the name of a character string and a list of arguments, assembles an expression giving the function call, and evaluates it. The function is quite simple in S (it is an internal function in R)

```
do.call <- function(what, args = list()) {
  if(!is.list(args)) stop("args should be a list")
  if(length(args)) this.call <- c(as.name(what), args)
  else this.call <- list(as.name(what))
  mode(this.call) <- "call"
  eval(this.call, local = sys.parent(1))
}
```

and indicates how we might do the same thing 'by hand'. What is needed is an expression of mode `call` to be evaluated, and this can be assembled as a list with the first element the function to be called (of mode `name`) followed by the arguments. Named components of the list become specified arguments, and unnamed components become unspecified arguments. Once the list is assembled, it is given mode `"call"`. (It is normally inadvisable to change modes in this way, and it may be safer to use `as.call`.)

If only the function to be called is to be decided at run time, `call` can be used, which takes as first argument a character string giving the function name followed by the arguments to be used. It is equivalent in S to the function

```
mycall <- function(NAME, ...)
{ Call <- sys.call(); Call[[1]] <- as.name(NAME); Call }
```

Note that the list of arguments of a function call really is a list, so if we want to alter one element as here we use `[[ ]]`.

Very much the same ideas can be used to manipulate an existing call, treating it as a list. For example, in our library `spatial` we have

```
surf.gls <- function(np, covmod, x, y, z, nx = 1000, ...)
    ....
covmod <- covmod # needed in new S engine
args <- list(...)
if(length(args)) {
  pm <- pmatch(names(args), names(covmod), nomatch=0)
  if(any(pm == 0)) warning(paste("some of ... do not match"))
  covmod[pm] <- unlist(args)
}
```

which adds any arguments given in `...` as arguments to the function `covmod`. (This modified function is returned in the fit object, which is why we cannot just use `covmod(x, ...)` as usual.) Here we rely on the names of a function being the names of the arguments, as the final list component, the function body, is unnamed.

We can now explain more fully how `args` works. The version for the new S engine is

```
args <- function(x) {
  if(mode(x) != "function") {
    if(mode(x) == "character") x <- getFunction(x)
    else stop("need the name of a function")
  }
  x[length(x)] <- function() NULL
  x
}
```

The initial part enables us to use `args("hist")` or even `args("[<-.factor")`. The meat of the function sets the body of the function to be empty.

Functions in R have to be treated somewhat differently: the R version of our    R
snippet from `surf.gls` is

```
if(length(args)) {
  oargs <- formals(covmod); onames <- names(oargs)
  pm <- pmatch(names(args), onames, nomatch = 0)
  if(any(pm == 0)) warning(paste("some of ... do not match"))
  names(args[pm > 0]) <- onames[pm]
  formals(covmod)[pm] <- args[pm > 0]
}
```

We use `formals` to extract a named list of the arguments and their defaults,
and later as a replacement function to replace the defaults (and we cannot rely
on list replacement to extract the elements we need from the right hand side).
If we need the function body, this is given (or replaced) by a call to `body`, and
is again of mode `call`. One useful function in connection with `formals` is
`alist` which is like `list` but allows named components with empty values (as
in `alist(newarg=)`, adding an argument with no default).

R       Function `args` in R is an internal function, but it could have been written as

```
myargs <- function(x) {
  if(is.function(x)) x <- match.fun(x)
  body(x) <- list(NULL)
  environment(x) <- .GlobalEnv
  x
}
```

Note that as `x` becomes a new function defined within `myargs`, it will acquire
as enclosing environment the body of `myargs`. Here that is undesirable (printing
the function would give the address of the environment), so we reset it.

### Manipulating unevaluated expressions

Sometimes it is desirable or even necessary to keep expressions unevaluated. The
new S engine provides the function `Quote` for this purpose: it takes a single
argument and passes it on, bypassing the evaluation mechanism. In the other
engines `substitute(expr, list(NULL))` can be used for this purpose, and R
has the equivalent `quote`. These constructions are not very common, although
they can be used in the S engines for computing calls for use as the `tracer`
argument of `trace` (see page 185).

More commonly `substitute` is used to substitute values in the expression
while leaving it unevaluated. Without a second argument it replaces any occurrences of variables in the local frame (notably arguments which are treated slightly
differently in S and R: see page 44) by their values: the second argument can be
a list or frame to give `name` = `value` pairs to be substituted. We have already
seen the use of `substitute` to produce labels. Our function `family.negbin`
in library MASS provides another example:

```
family.negbin <- function(object)
  eval(substitute(
    negative.binomial(theta = object$theta, link = lnk),
    list(lnk = object$call$link)
  ))
```

This extracts the family as understood by `glm` from an object of class `negbin`
by substituting the appropriate values of `theta` and `link`. (This is a little simpler than using `paste` and `parse`.) Actually, in this example the substitution is
needed only because `negative.binomial` contains the line

```
link <- as.character(substitute(link))
```

to allow the usage `negative.binomial(2, log)` rather than `"log"`. This follows the usage of `binomial` and `poisson`, but it would be better to use

```
if(mode(link) != "character" && mode(link) != "name")
    link <- as.character(substitute(link))
```

## Evaluating in other frames

The real power of computing on the language comes from the ability to evaluate expressions in other frames. Function `eval` in the S engines has arguments

```
eval(expression, local = T, parent)
```

The first argument is evaluated (as a language object: that is any lazy evaluation that has been postponed is now done) in the local frame, and the language object is then evaluated in the frame given by the second argument, by default the frame in which `eval` was called. Argument `local` can be any frame (by number or as a list) or if false the 'global frame' 0 is used. Argument `parent` is used only if `local` is a list (and is designed for system usages such as in browsers). A common choice for `local` is `sys.parent()`, which evaluates the expression in the frame of the caller of the current function.

In R function `eval` has arguments                                        R

```
eval(expr, envir = parent.frame(), enclos)
```

By default this has the same effect as S; it evaluates[12] the first argument as a language object and then evaluates it in the environment from which `eval` was called. (Function `eval.parent(expr, n)` provides a convenient shorthand for evaluating n environments up the call sequence.) The enclosing environment `enclos` is that to be used in applying R's scope rules. It is used only when `envir` is a list and sets the scope for items not found on the list.

We do have to be very careful in considering evaluation. Because of lazy evaluation, some S objects will automatically be evaluated in other frames, most notably actual function arguments, but this can be confusing. For example, a `predict` method might be written

```
predict.my <- function(object, newdata = object$call$data) {
    newdata <- eval(newdata, sys.parent())
    . . . .
```

It is tempting to think that `newdata` will be evaluated in the parent frame. That is true of an actual argument, but it is not true of the default expression, which in this case will evaluate to an object of mode `"name"` (if `object$call` is a matched call). Thus we can safely use the `eval` call given, although unless we are being perverse[13] any actual argument will already be evaluated.

---

[12] use `evalq` to avoid this.

[13] for example, supplying `newdata` of a language mode.

Sometimes our requirements are more complicated. We have noted that the first argument, `expression`, of `eval` *is* evaluated, so we may need to ensure that it is evaluated in the right place. Suppose we want to write a function to plot a curve specified by an expression. The first argument is to be an expression in `x` that evaluates to a vector of the same length of `x`. We need to use

```
curvePlot <- function(expr, xlim = c(0, 1)) { # incomplete
  x <- seq(xlim[1], xlim[2], len = 200)
  plot(x, eval(substitute(expr)), type="l")
}
curvePlot(sin(x), c(0, 2*pi))
```

as the expression must be evaluated in the local frame where `x` is defined. Without the `substitute` call, `expr` would be evaluated in the parent frame (as it is an argument) before being passed on to `eval`, and pick up the value of `x` (if any) visible from the parent frame. On the other hand, `substitute` is needed rather than `Quote`, for `Quote(expr)` is just `expr` of mode `name`, and we need to substitute to pick up the actual value of the argument `x`.

However, this function is still not precisely what we want, as `expr` might refer to other objects defined in the parent frame; for example

```
testit <- function() {
  omega <- 3.7
  curvePlot(sin(omega*x), c(0, 2*pi))
}
testit()
```

R
will fail. So we want to evaluate `expr` with `x` from the local frame and other objects from the parent frame. We can do this in the S engines by constructing an explicit frame within which to do the evaluation, starting with a copy of the parent frame.

```
curvePlot <- function(expr, xlim = c(0, 1)) {
  x <- seq(xlim[1], xlim[2], len = 200)
  lf <- sys.frame(sys.parent()); lf$x <- x
  plot(x, eval(substitute(expr), lf), type="l")
}
```

We could try copying only the variables in the parent frame that are needed to evaluate the expression, but it is hard to know what those are without evaluating the expression. Strategies such as putting `x` in the parent frame (this might overwrite a variable there) or making `x` globally visible (in frame 0 or 1, possibly masked by the evaluation frame) and then evaluating `expr` in the parent frame are prone to puzzling errors. The following may require less copying:

```
curvePlot <- function(expr, xlim = c(0, 1)) {
  x <- seq(xlim[1], xlim[2], len = 200)
  needCopy <- exists("x", frame = sys.parent())
  if(needCopy) xx <- get("x", frame = sys.parent())
  assign("x", x, frame = sys.parent())
```

```
        plot(x, expr, type="l")
        if(needCopy) assign("x", xx, frame = sys.parent())
}
```

Here we rely on lazy evaluation, as `expr` will not be evaluated (from within `plot`, but in the parent frame of `curvePlot`, as it is an actual argument) until we have replaced `x` in the parent frame.

In R we can look at the function `curve` for inspiration. The idea is to use the    R `enclos` argument of `eval`:

```
curvePlot <- function(expr, xlim = c(0, 1)) { # R only
    x <- seq(xlim[1], xlim[2], len = 200)
    plot(x, eval(substitute(expr), list(x=x),
                sys.frame(sys.parent())), type="l")
}
```

This works because the search path for objects in `expr` is first the list, then the search path from the caller of `curvePlot`.

For another example, consider the code[14] used in `predict.lda`

```
if(!is.null(sub <- object$call$subset))
    newdata <- eval(parse(text = paste(deparse(object$call$x),
                "[", deparse(sub), ",]")), sys.parent())
else newdata <- eval(object$call$x, sys.parent())
if(!is.null(nas <- object$call$na.action))
    newdata <- eval(call(nas, newdata))
```

to retrieve the data used in the original fit if argument `newdata` is missing. Here the effect is to evaluate the data in the frame of the caller. This example has most of the features of computing on the language that we have covered. We can test this by an example from Chapter 11 of MASS, embedded in a function and extended to test some extra features. (Use `iris3` in R.)

```
test <- function() {
    ir <- rbind(iris[,,1], iris[,,2], iris[,,3])
    ir.species <- factor(c(rep("s",50), rep("c",50), rep("v",50)))
    ir.lda <- lda(log(ir), ir.species, na.action=na.omit,
                subset = ir.species != "v")
    predict(ir.lda, dimen=2)$x
}
test()
```

Normally we use `eval(object$call$x, sys.parent())`; the first argument is evaluated in the local frame where `object` is defined, and so `object$call$x` is evaluated to `log(ir)`. This is then evaluated in the parent frame where `ir` is visible. We modify the expression in the case of a `subset` argument: we have to be careful to keep that unevaluated as it needs to be evaluated in the parent frame.

---

[14]replace sys.parent() by sys.frame(sys.parent()) in R.

## 3.6   Graphics functions

Trellis graphics in the S engines is quite programmable, although it can be less than obvious how to do so. We give in MASS some examples[15] of the use of panel functions that can be used to customize what is plotted on each panel of, for example, an `xyplot` call. A panel function generally has arguments `x`, `y` and `...`, the latter being used for graphical parameters that are passed down. Here `x` and `y` are the coordinates of the points to be plotted in that panel.

There is a fairly conventional style for panel functions that will pick up the Trellis default styles for the current device, For example, `panel.loess` is

```
function(x, y, span = 2/3, degree = 1,
    family = c("symmetric", "gaussian"), evaluation = 50,
    lwd = add.line$lwd, lty = add.line$lty, col = add.line$col,
    ...)
{
    add.line <- trellis.par.get("add.line")
    lines(loess.smooth(x, y, span = span, degree = degree,
                    family = family, evaluation = evaluation),
            lwd = lwd, lty = lty, col = col, ...)
}
```

If a more detailed view of the data is needed it may be essential to know which parts of the data frame have been mapped into that panel. This is the purpose of the argument `subscripts`, which gives the indices of the selected points. We could use this to, for example, name the points on the plot or select a colour based on the value of another observation. The `subscripts` argument is used by `panel.levelplot` to pick out the `z` coordinates of the points that are based to that panel: this can be seen in the following example from MASS page 81

```
levelplot(pred ~ x * y, topo.plt, aspect=1,
    at = seq(690, 960, 10), xlab="", ylab="",
    panel = function(x, y, subscripts, ...) {
        panel.levelplot(x, y, subscripts, ...)
        panel.xyplot(topo$x,topo$y, cex=0.5, col=1)
    }
)
```

where `z` is passed as part of the `....`.

Panel functions can also be used with some of the traditional plot functions, for example the default method of `pairs` and `coplot`. See `pairs.lda` in our library MASS for one example.

Trellis also has prepanel functions, which are used to help it in layout and scaling. MASS has one example on page 115:

---

[15]For example, pages 62, 78–82, 84, 87, 115, 150, 200, 250, 266, 277.

```
x <- rt(250, 9)
qqmath(~ x, distribution=qnorm, aspect="xy",
    prepanel = prepanel.qqmathline,
    panel = function(x, y, ...) {
        panel.qqmathline(y, distribution=qnorm, ...)
        panel.qqmath(x, y, ...)
    },
    xlab = "Quantiles of Standard Normal"
)
```

The purpose of a prepanel function is to indicate the $x$ and $y$ limits of the plot region, and also to help choose the aspect ratio if aspect = "xy" is used, by giving values dx and dy that need to be equal distances on the $x$ and $y$ axes. The system version of prepanel.qqmathline is not quite right: we think it should be

```
function(x, y, distribution)
{
    xlim <- range(x); ylim <- range(y); dx <- diff(xlim)
    y <- quantile(y, c(0.25, 0.75))
    x <- distribution(c(0.25, 0.75))
    slope <- (y[2] - y[1])/(x[2] - x[1])
    intercept <- y[2] - x[2] * slope
    ylim <- range(slope * xlim + intercept, ylim)
    list(xlim = xlim, ylim = ylim, dx = dx, dy = slope*dx)
}
```

so it is the 'null' line that is banked at $45°$.

# Chapter 4

# Classes

This chapter covers the class-oriented features of the old S engine and of R. Most of these exist for backward compatibility in the new S engine (see Section 5.4), but the features introduced in the next chapter are preferred for new projects in that system.

S is a functional language and so is class-oriented in a somewhat different sense from C++, Java or Visual Basic: rather than data objects having methods, functions have methods that behave differently for different classes of data objects. Crucially, the set of methods is dynamic and users can add new methods to existing functions.

## 4.1    Introduction to classes

The primary purpose of the S programming environment is to construct and manipulate objects. These objects may be fairly simple, such as numeric vectors, factors, arrays or data frames, or reasonably complex such as an object conveying the results of a model fitting process. The manipulations fall naturally into broad categories such as plotting, printing, summarizing and so forth. They may also involve several objects such as performing an arithmetic operation on two objects to construct a third.

Since S is intended to be an extensible environment new kinds of object are designed by users to fill new needs, but it will usually be convenient to manipulate the objects using familiar functions such as `plot`, `print` and `summary`. For the new kind of object the standard manipulations will usually have an obvious purpose, even though the precise action required differs at least in detail from any previous action in this category.

Consider the `print` function. It is normally invoked implicitly simply by evaluating an expression, or giving a name, at the session level, as in

```
> x <- 1:5
> x
[1] 1 2 3 4 5
```

If a user designs a new kind of object called, say, a "`newfit`" object, it will often be useful to make available a method of printing such objects so that the important features are easy to appreciate and the less important details are suppressed.

One way of doing this is to write a new function, say `print.newfit`, which could be used to perform the particular kind of printing action needed. If `myobj` is a particular `newfit` object it could then be printed using

```
> print.newfit(myobj)
```

We could also write functions with names `plot.newfit`, `summary.newfit`, `residuals.newfit`, `coefficients.newfit` for the particular actions appropriate for `newfit` objects for plotting, summarising, extracting residuals and extracting coefficients, and so on.

In the case of printing, however, it would be much more convenient for the user if simply naming the object at the session level

```
> myobj
```

were enough to cause the special printing action to happen. Similarly it would be useful to be able to use the standard functions such as `plot(myobj)` and `summary(myobj)` and have the S evaluator select the appropriate action for the kind of object presented to it.

This is one important idea behind *object-oriented programming*. To make it work we need to have a standard method by which the evaluator may recognise the different kinds of object being presented to it. In the old S engine and in R this is done by giving the object a `class` attribute, which at its simplest is a character string naming the class. There is a `class` replacement function available for making this attribute assignment, which would normally be done when the object was created. For example

```
> class(myobj) <- "newfit"
```

In some cases the `class` attribute is a character vector of several components; the object is then said to *belong* to the class specified by the first and to *inherit* from classes specified by the second, third, and so on, in turn.

The function `unclass` removes the class(es) of its argument.

## Generic and Method Functions

Functions like `print`, `plot` and `summary` are called *generic* functions. They have the property of adapting their action to match the class of object presented to them. Typically their definition is very short, often only one line:

```
> print
function(x, ...)
UseMethod("print")
```

The UseMethod function is treated specially by the evaluator. The first step taken is to evaluate the principal argument (here x) and examine its class attribute, if any. If the object x belongs to class newfit, and if there is a function print.newfit visible on the search path, then the remaining body of the generic function (usually, as here, all of it) is replaced by a call to print.newfit of the form print.newfit(x, ...), but using the existing evaluation frame rather than generating a new frame and leaving the remaining arguments unevaluated. Objects created in the frame before the call to UseMethod will remain visible unless masked by an argument of the same name. In addition the frame at this point will always contain a few extra objects named .Class, .Method and .Generic (and .Group in the S engines). These are character strings giving information on the process underway. Any code in the generic function that would normally be executed after the call to UseMethod is (silently) discarded.

Functions such as print.newfit are called *method* functions for the print generic function. The entire process is called *method dispatch*.

If no function print.newfit is visible the class vector of x is examined to see if there is a print method function available for one of the classes from which the object inherits. If so, the first one found is used. If there is no primary or inherited print method function available, or if the object x has no class attribute, then the default method, here print.default, is used. If no method function can be found for a generic function and no default method is supplied an error results.

The advantages of this approach include being able to add methods to existing generic functions for new classes without any need to change existing code, being able to manipulate objects using a consistent and familiar suite of generic functions, and through inheritance being able to capitalise on existing code when some existing method caters for objects in the new class sufficiently well. The scheme offers most advantage to the programmer if the classes can be built incrementally so that maximum use may be made of the inheritance mechanism to re-use existing code.

There is an apparently undocumented trap in writing method functions.[1] The principal argument of the method function must have the same name as the principal argument of the generic function; thus the first argument of all print methods should be x, and of all summary, predict, coef, ... methods should be object. Otherwise the principal argument will be evaluated twice, once by the generic function and once by the method function. If the first argument is a call to a fitting function, the (possibly time-consuming) fit would be performed twice.

*Explicitly invoking inheritance*

This and the next subsection describe tricky aspects of method dispatch that are best avoided if possible.

The function NextMethod function can be used within methods to call the next method. That is, it looks down the list of classes starting with the one after the

---

[1] at least in the S engines.

one that caused the current method to be invoked, and looks at each in turn to see if there is a method for the current generic, calling the first method found and if none calling the default method if one exists. Unlike UseMethod, NextMethod does return to the calling function.

It is not entirely clear what should happen if no method is found or if the list of classes has been changed since the current method was selected. In the S engines the following example will alternate classes c1 and c2 until it hits the limit on nesting expressions, rather than report that there is no method to invoke (which is what R does).

R

```
test <- function(x) UseMethod("test")
test.c1 <- function(x) {cat("c1\n"); NextMethod(); x}
test.c2 <- function(x) {cat("c2\n"); NextMethod(); x}
test.c3 <- function(x) {cat("c3\n"); x}
x <- 1; class(x) <- c("c1", "c2"); test(x)
```

If we add a default method by

```
test.default <- function(x) {cat("default\n"); x}
```

all the engines call c1, c2 and then the default. On the other hand, if we use

```
test.c1 <- function(x)
   {cat("c1\n"); class(x) <- "c3"; NextMethod(); x}
x <- 1; class(x) <- c("c1", "c2"); test(x)
```

the S engines call c1, c2 and then the default if one is present, otherwise c3! Such constructions are best avoided.

It is possible to call NextMethod with arguments, but what happens is engine-dependent. We recommend calling the generic with the new arguments instead.

### Dispatching on other arguments

Method dispatch does not have to be based on the first argument: using UseMethod("loglm", data) causes the class of data (the second argument) to be used. (That is why we carefully said 'principal' argument up to now.)

There are a small number of functions that select a method to dispatch based on more than one argument. An example is cbind that uses the classes of all of the ... arguments to select a suitable method. Precisely how the class is chosen seems undocumented, but the effect is to choose a method function that applies to all the arguments that do have classes, or use the default.

The standard binary operators are also generic and have special dispatch rules in the S engines. If just one argument has a class with an applicable method it is used for method dispatch. If both do and they are the same, that method is used; otherwise the default method is used with a warning.

## A generic $t$-test function

We illustrate some of the ideas of object-oriented programming by revisiting the $t$-test problem discussed on pages 42 and 45. We will allow the two samples to be specified in a number of different ways, namely

1. as two separate numeric vectors, as in the previous case,
2. as a two-column matrix,
3. as a list of two components, or
4. as a two-level factor and a single numeric vector.

The first argument will determine the particular method dispatched, but since numeric vectors, matrices and lists normally will have no class attribute, we need to assign one to them so that the method dispatch will be appropriate. The function `data.class` infers a class from attributes of the object. If a class attribute exists the inferred class is its first element (as a character vector). In other cases inferred classes often coincide with the mode of the object, such as `"numeric"` or `"list"`, but for matrices the inferred class is `"matrix"`. For our purposes it is important to know that our four cases will be distinguished.

As usual the generic function is very simple:

```
Ttest <- function(z, ...) {
  if(is.null(class(z))) class(z) <- data.class(z)
  UseMethod("Ttest")     # UseMethod("Ttest", z, ...) in R
}
```

Since `data.class` does not preserve having more than one class it can be important not to use it when a class for the object already exists.

Note that this generic function may change the argument `z`. The implementors of S and of R differ on how that should be interpreted (that is, in their reading R of Chambers & Hastie, 1992, §A.6).

Next, consider a default method. This is almost the same as the previous `ttest` function with the original samples and arguments returned as part of the result.

```
Ttest.default <- function(z, y2, ..., alpha = 1/20,
        test = c("two-sided", "lower tailed", "upper tailed"))
{
  n1 <- length(z)
  n2 <- length(y2)
  ndf <- n1 + n2 - 2
  s2 <- ((n1 - 1) * var(z) + (n2 - 1) * var(y2))/ndf
  tstat <- (mean(z) - mean(y2))/sqrt(s2 * (1/n1 + 1/n2))
  tail.area <- switch(test <- match.arg(test),
    "two-sided" = 2 * (1 - pt(abs(tstat), ndf)),
    "lower tailed" = pt(tstat, ndf),
    "upper tailed" = 1 - pt(tstat, ndf),
    {
      warning("test must be 'two-sided', 'lower' or 'upper'")
```

```
        NULL
    })
res <- list("t-stat"=tstat, d.f.=ndf, y1=z, y2=y2,
            test=test, tail.area=tail.area,
            reject=tail.area < alpha, alpha=alpha)
class(res) <- "my.t.test"
res
}
```

The arguments `alpha` and `test` occur after a `...` argument which means they must be named, in full, if they are supplied by the user.

Consider now a print method for objects of this class. The key components are the first two, so a succinct `print` method simply prints those out.

```
print.my.t.test <- function(x, ...) {
  print(unlist(x[1:2]))
  invisible(x)
}
```

By convention `print` methods return the value of their principal argument invisibly. The `invisible` function turns off automatic printing, thus preventing an infinite recursion when printing is done implicitly at the session level.

We can now test the function with the two samples generated previously and shown on page 43:

```
> Ttest(x1, x2)
   t-stat d.f.
  -3.6303   18
```

If on some occasion we were only interested in the result and not the statistic, that component can still be accessed directly, as in

```
if(Ttest(x1, x2, test="upper")$reject) cat("x2 is better\n")
```

For most purposes the standard `print` method will be sufficient output, but if a more extensive display is required it can be done with a `summary` method. Most `summary` methods merely enhance the object with further components found by additional calculations, provide a new class and the display is done by a new print method. In this case the calculations are done already, so we merely change the class attribute and write a `print` method for the new class:

```
summary.my.t.test <- function(object)
  structure(object, class = c("sum.my.t.test", class(object)))

print.sum.my.t.test <- function(x, ...) {
  n1 <- length(x$y1); n2 <- length(x$y2)
  cat("\n T-test of  samples: \n")
  cat("Sample 1:", format(x$y1[1:min(n1, 5)], ...))
  cat(if(n1 <= 5) "" else " ...", "\n")
  cat("Sample 2:", format(x$y2[1:min(n2, 5)], ...))
```

```
      cat(if(n2 <= 5) "" else " ...", "\n")
      cat("\nt =", format(x$"t-stat", ...), "on",
          x$d.f., "d.f.\n")
      cat("Test:", x$test, "\n")
      cat("Tail area:", format(x$tail.area, ...), "\n")
      cat("Level:", format(x$alpha, ...), "\t")
      cat("Null hypothesis",
          if(x$reject) "rejected." else "retained.", "\n")
      invisible(x)
   }
```

The result is much more detail, including the first few sample members:

```
> tst <- Ttest(x1, x2, test = "lower")
> summary(tst)

 T-test of  samples:
Sample 1:   0.5 -1.4   1.5   0.5 -0.9 ...
Sample 2:   0.4 -0.3   0.9   2.2   1.8 ...

t = -3.6303 on 18 d.f.
Test: lower tailed
Tail area: 0.00095688
Level: 0.05       Null hypothesis rejected.
```

We can now write a `plot` method for `my.t.test` object. The idea is very simple. We simply extract the sample components from the object and call `boxplot`:

```
plot.my.t.test <- function(x, ...)
   invisible(boxplot(x[c("y1", "y2")], ...))
```

Any customisation needed may be supplied as additional arguments which are passed on to `boxplot`. The actual call is to `plot`, as in

```
tst <- Ttest(x1, x2)
plot(tst)
```

Finally consider the alternative ways we wish to allow for calling `Ttest`. These are handled by very short method functions that take the initial argument or arguments, manufacture arguments that suit `Ttest.default` and effectively call that function.

```
Ttest.matrix <- function(z, ...) Ttest(z[, 1], z[, 2])
Ttest.list <- function(z, ...) Ttest(z[[1]], z[[2]])
Ttest.factor <- function(z, y, ...) {
   lev <- levels(z)
   Ttest(y[z == lev[1]], y[z == lev[2]])
}
```

A test of the `factor` method is

```
> x12 <- c(x1, x2)
> sam <- factor(rep(0:1, c(length(x1), length(x2))))
> Ttest(sam, x12)
  t-stat d.f.
 -3.6303    18
```

Our libraries contain many examples of the use of the object-oriented programming paradigm. Most often we just add method functions for generic functions such as print, summary and predict, as in classes "lda", "qda" and "nnet". Similarly objects of our class "negbin" inherit from class "glm" but need their own method functions for anova, summary and family to cater for the few special needs of negative binomial models.

Our robust linear regression functions discussed in MASS Section 6.5 define a new class "rlm", and also inherit from the linear regression class "lm". Both print and summary functions are provided. However, summary.rlm produces a result of class "summary.lm", so printing the summary invokes print.summary.lm.

In general inheritance works well only when the classes are designed in parallel with inheritance in mind. For example, it might appear that we could use inheritance from summary.lm to build summary.rlm, but scale estimation by the root mean square of the residuals is so embedded in summary.lm that this proves to be impossible.

## relevel: a generic utility

We saw on page 14 that some statistical functions give special weight to the first level of a factor, so it is convenient to have a function that makes a specified level the first one. This is the purpose of BDR's function relevel, now incorporated into R. We want it to apply only to factors, and we use the method dispatch mechanism to ensure this. The generic function is

```
relevel <- function (x, ref, ...) UseMethod("relevel")
```

with default method

```
relevel.default <- function (x, ref, ...)
  stop("relevel only for factors")
```

The method for factors is quite short

```
relevel.factor <- function (x, ref, ...)
{
  lev <- levels(x)
  if(is.character(ref)) ref <- match(ref, lev)
  if(is.na(ref)) stop("ref must be an existing level")
  nlev <- length(x)
  if(ref < 1 || ref > nlev)
    stop(paste("ref =", ref, "must be in 1 :", nlev))
  factor(x, levels = lev[c(ref, seq(along = lev)[-ref])])
}
```

and this looks to be all we need to do. Unfortunately, not so. Although ordered factors appear to be a special case of factors and so have been given class

```
c("ordered", "factor")
```

inheriting from `"factor"`, in this respect the statistical methods do *not* treat them like factors. So we need to override the behaviour of `relevel` for ordered factors, by

```
relevel.ordered <- function (x, ref, ...)
    stop("relevel only for factors")
```

This illustrates the treacherous nature of inheritance.

## 4.2   An extended statistical example

Model formulae have become much more widely used in S in recent years. We will explore the paradigms of using model formulae via our `lda` function (for linear discriminant analysis), which was originally written for S version 2, and was a single function in the first edition of MASS. For the second edition we made the function generic, renaming the workhorse function as `lda.default`. Thus we have

```
lda <- function(x, ...)
{
   if(is.null(class(x))) class(x) <- data.class(x)
   UseMethod("lda")
}
lda.default <- function(x, grouping, prior = proportions,
   tol = 0.0001, method = c("moment", "mle", "mve", "t"), CV=F,
   nu = 5, ...)
      ....
```

(The full sources of these functions are in library MASS.) Note that the generic function may change the argument x. So in R we need                                    R

```
lda <- function(x, ...)
{
   if(is.null(class(x))) class(x) <- data.class(x)
   UseMethod("lda", x, ...)
}
```

That ought to work in S too, but crashes under at least one implementation.

When working with model formulae it is helpful to store the function call as component `call` of the output list. There is a standard way to do this using the function `match.call`, so the return value of `lda.default` becomes

```
structure(list(prior = prior, counts = counts, means =
             group.means, scaling = scaling, lev = levels(g),
             svd = X.s$d[1:rank], N = n, call = match.call()),
          class = "lda")
```

Once this is done the function `update` can be used to update the object just as for linear models. It is also useful to have a record in the object of how it was created. Note that it is not possible to record the call of the generic function, as `UseMethod` does not return but merely implements method dispatch.

We allow the first argument of `lda` to be either a matrix-like object (such as a data frame) or a formula. Since matrices do not necessarily have a class assigned, our generic function `lda` assigns a class via the function `data.class`. The method despatch mechanism will then use the functions `lda.matrix`, `lda.data.frame` and `lda.formula` as appropriate.

The function `lda.data.frame` is quite simple, as it calls the matrix method and then fixes up the recorded call

```
lda.data.frame <- function(x, ...)
{
  res <- lda.matrix(data.matrix(x), ...)
  res$call <- match.call()
  res
}
```

It is tempting to use `NextMethod`, but this calls the *second* method, here the default method.

We could do without an explicit `lda.matrix` and rely on `lda.default`, but we chose to add `subset` and `na.action` arguments for matrices.

```
lda.matrix <- function(x, grouping, ...,
                       subset, na.action = na.fail)
{
  if(!missing(subset)) {
    x <- x[subset, , drop = F]
    grouping <- grouping[subset]
  }
  if(!missing(na.action)) {
    dfr <- na.action(structure(list(g = grouping, x = x),
                               class = "data.frame"))
    grouping <- dfr$g; x <- dfr$x
  }
  res <- NextMethod("lda")
  res$call <- match.call()
  res
}
```

After these preliminaries we can write the method for model formulae. How would we wish to interpret a model formula? Our idea was to follow the closely related logistic discrimination, so the left-hand side should be a factor giving which group the case belongs to, and the right-hand side should give a vector of explanatory variables. We do *not* want an intercept term (LDA centres groups on the group mean), so any intercept term (usually implicit) should be ignored. It would not be usual to expand factors or to allow interactions, but there seems no reason to restrict a knowledgeable user. It should come as no surprise that we construct a matrix and grouping factor, call the default method and adjust the recorded call.

```
lda.formula <- function(formula, data = NULL, ...,
                        subset, na.action = na.fail)
{
  m <- match.call(expand.dots = F)
  if(is.matrix(eval(m$data, sys.parent())))
      m$data <- as.data.frame(data)
  m$... <- NULL
  m[[1]] <- as.name("model.frame")
  m <- eval(m, sys.parent())
  Terms <- attr(m, "terms")
  grouping <- model.extract(m, "response")
  x <- model.matrix(Terms, m)
  xint <- match("(Intercept)", dimnames(x)[[2]], nomatch=0)
  if(xint > 0) x <- x[, -xint, drop=F]
  res <- lda.default(x, grouping, ...)
  res$terms <- Terms
  res$call <- match.call()
  res
}
```

The first steps are to match the call, without expanding the ... term, and then to ensure that the `data` argument is a list or data frame, not a matrix. We then call `model.frame` with the essential subset of the arguments of the original call, setting all arguments that do not apply to `model.frame` to `NULL`. (At this level of abstraction all the arguments specific to `lda` are still in the ... term.) The call to `model.frame` returns a data frame with columns corresponding to the response, the main effects and any extra variables such as weights, handling `na.action` and `subset`. If factors are not expected, it is simpler to remove the intercept in the formula by

```
attr(Terms, "intercept") <- 0
x <- model.matrix(Terms, m)
```

but this does not do the right thing when factors *are* used.

Note that `lda.formula` breaks one of the rules: its first argument should be called `x` to avoid it being evaluated twice. Unfortunately, it also needs to be called `formula` for `update` to work properly. We *could* get around that by writing a method for `update`, but evaluating a formula twice is normally a negligible cost. If we were starting afresh, we would probably call the principal argument `formula`, not `x`.

### The `predict` method

For the `predict` method we have to be able to extract the data matrix `x` from a new data frame. We made life easy by storing the `terms` attribute in the `lda` object, so we can use the following function. (This has become much more complicated with every new version!)

```
predict.lda <-
  function(object, newdata, prior = object$prior, dimen,
  method = c("plug-in", "predictive", "debiased"), ...)
    ....
  if(!is.null(Terms <- object$terms)) {
    # formula fit
    if(missing(newdata)) newdata <- model.frame(object)
    else {
        newdata <- model.frame(as.formula(delete.response(Terms)),
                                  newdata, na.action=function(x) x)
    }
    x <- model.matrix(delete.response(Terms), newdata,
                        contrasts = object$contrasts)
    xint <- match("(Intercept)", dimnames(x)[[2]], nomatch=0)
    if(xint > 0) x <- x[, -xint, drop=F]
  } else {
    # matrix or data-frame fit
    if(missing(newdata)) {
      if(!is.null(sub <- object$call$subset))
        newdata <- eval(parse(text=paste(deparse(object$call$x),
                          "[", deparse(sub),",]")), sys.parent())
      else newdata <- eval(object$call$x, sys.parent())
        if(!is.null(nas <- object$call$na.action))
          newdata <- eval(call(nas, newdata))
    }
    if(is.null(dim(newdata)))       # a row vector
      dim(newdata) <- c(1, length(newdata))
    x <- as.matrix(newdata)  # to cope with data frames
  }
    ....
```

This has to handle both cases in which newdata is a new matrix-like object and those in which it is a data frame from which the matrix is to be extracted. We allow the first possibility only if the original call specified a matrix-like object and the second only if a formula was given. The purpose of the rest of the code is to attempt to retrieve the original data in the parent frame if newdata is not specified.

Some of the details of predict.lda may not be obvious. We must not assume that the object was created by lda, so we identify formula-based objects by the presence of the terms component. Asking for

```
model.matrix(Terms, newdata)
```

will cause an error unless the left-hand side of the formula can be matched; delete.response works around this. For details of the evaluation of a missing newdata argument see page 71.

Eventually we added a method for model.frame: it is used by methods for the standard generic functions to manipulate model fits, for example predict.

```
model.frame.lda <-
function(formula, data = NULL, na.action = NULL, ...)
{
  oc <- formula$call
  names(oc)[2:3] <- c("formula", "data")
  oc$prior <- oc$tol <- oc$method <- oc$CV <- oc$nu <- NULL
  oc[[1]] <- as.name("model.frame")
  if(length(data)) {
    oc$data <- substitute(data)
    eval(oc, sys.parent())
  } else eval(oc, list())
}
```

It was not necessary to use the generic function mechanism to use model formulae with `lda`; the function `princomp` in S-PLUS shows a different approach.
However, the class mechanism is powerful and there to be used.

## 4.3 Polynomials: an example of group method functions

The unary and binary operators such as `+`, `-` and `>` are functions in S: an expression such as `x + 3` is merely a shorthand for `"+"(x, 3)` which is also a
legitimate usage.

The operators are generic functions, so methods may be written for them.
They are special for two reasons. First, method dispatch is determined by the class
of both arguments (See page 78.) Second, it is possible to write a single *group*
method function that provides methods for all the arithmetic and logical operators
in one function. In some languages providing methods for the operators is called
"operator overloading" but in S operator methods do not differ in principle from
other method functions.

### Polynomial objects

We will illustrate methods for the operators by constructing a class[2] for polynomials of one variable, which we shall simply think of as $x$. Polynomial operations
like addition, subtraction and multiplication are familiar and obey very similar
rules to those of ordinary arithmetic. Our primary aim will be to provide facilities
in S to allow these operations in a syntactically familiar way. This will achieve
more than mere convenience. If `p1`, `p2` and `p3` are polynomial objects in S
writing

```
P <- p1 + p2*p3
```

for an operation that otherwise might be written

```
P <- poly.add(p1, poly.mult(p2, p3))
```

---

[2] Only some of the methods are shown here: the on-line script for this chapter contains all the code
and several additional methods.

is not only clearer and simpler but much less error-prone. The first step is to decide on what a polynomial object such as p1 will look like in S.

In constructing a class it pays to consider carefully how its objects should be represented. The most delicate issue is often the balance between simplicity and generality. For many reasons it seems sensible to limit ourselves to polynomials in one generic variable here. This greatly simplifies our task, but we should be aware that if ever we need to extend to, say, bivariate polynomials we may need to re-design the class entirely.

A polynomial object will be represented here by a vector of numeric coefficients with a class attribute. They will be written in 'power series' form, that is with the powers in increasing order, and the leading (last) coefficient will be guaranteed to be non-zero unless the polynomial itself is identically zero. A suitable constructor function for polynomial objects is

```
polynomial <- function(a = c(0, 1)) {
  if(!(is.numeric(a) || is.complex(a))) {
    warning("non-numeric coefficients")
  }
  a <- as.numeric(a)
  if(any(!is.finite(a))) stop("invalid coefficient vector")
  la <- la0 <- length(a)
  if(la == 0) a <- 0
  else {
    while (a[la] == 0 && la > 1) la <- la - 1
    if(la < la0) a <- a[1:la]
  }
  class(a) <- "polynomial"
  a
}
```

The first step coerces a to be numeric, but just as importantly strips off other attributes such as dim, dimnames or names which if present might cause problems later. The next step ensures the leading coefficient is non-zero (unless the polynomial is the zero polynomial). Note that zero coefficients of lower degree are not omitted so the degree of the polynomial is always one less than the length of the coefficient vector, with again the exception of the zero polynomial. Adopting the power series form ensures that the coefficient of $x^k$ is a[k+1] .

With any class it is conventional and often useful to have two other functions defined, a *predicate* ( is.xxx ) function and a *coercion* ( as.xxx ) function. For polynomials these are very simple

```
is.polynomial <- function(p) inherits(p, "polynomial")

as.polynomial <- function(p)
  if(is.polynomial(p)) p else polynomial(p)
```

### Extracting or replacing coefficients

So much of S relies on indexing that defining a subset method for a vector-like class is usually a very early step. In this case we only need methods for `"["` and `"[<-"`.

```
"[.polynomial" <- function (x, i, ...) {
    x <- unclass(x)
    polynomial(NextMethod("["))
}

"[<-.polynomial" <- function (x, i, value) {
    value <- as.numeric(value)
    x <- unclass(x)
    x[i] <- value
    polynomial(x)
}
```

These will behave as operations on the coefficient vector, re-casting to a polynomial afterwards. Other conventions are possible but it is not obvious what should be done if, for instance, the same coefficient is selected twice in the operation.

For list-like classes we would usually need methods for `"[["`, `"[[<-"`, `"$"` and `"$<-"`.

### Unary and binary operations

Methods for the unary and binary operators may be written the same way any method functions are written. For example subtraction may be allowed by writing a method function such as

```
"-.polynomial" <- function(e1, e2)
{
    if(missing(e2)) return(polynomial(-unclass(e1)))
    e1 <- c(unclass(e1), rep(0, max(0, length(e2) - length(e1))))
    e2 <- c(unclass(e2), rep(0, max(0, length(e1) - length(e2))))
    polynomial(e1 - e2)
}
```

Notice that with the unary form of the operator as in `p2 <- -p1` the missing argument is `e2`, not `e1`.

The addition method could be almost identical but rather than work our way tediously through all the operators it is preferable to write a single method function for the *group generic function* `Ops`. The method dispatch mechanism will then call the function `Ops.polynomial` when a polynomial method is needed for any operator, supplying the name of the operator that initiated the dispatch as a local frame variable, `.Generic`. Group method functions usually have a `switch` statement using the `.Generic` variable to separate the specific actions required.

| Group | Component functions |
|-------|---------------------|
| Ops | +, -, *, /, ^, %/%, %%, <, <=, >, >=, ==, !=, &, \|, ! |
| Math | abs, acos, acosh, asin, asinh, atan, atanh, ceiling, cos, cosh, cumsum, exp, floor, gamma, lgamma, log, log10, round, signif, sin, sinh, tan, tanh, trunc |
| Summary | all, any, max, min, prod, range, sum |

**Table 4.1**: The group generic functions and their components.

Table 4.1 lists the group generic functions, together with the component functions that initiate method dispatch to each of them. Note that %*% is a generic binary operator but not a member of a group and %o% is not generic.

Note that the three group generic functions Ops, Math and Summary are *not* visible objects in S but may only be called indirectly through their component functions. Method functions written for them, such as Ops.polynomial, are necessarily visible but are also never called directly by the user. It is possible to write methods for individual members of a group, and these will take precedence over the group method.

Figure 4.1 gives a group method function for the unary and binary operators (of class Ops) which are useful for polynomials. The multiplication and division code relies on functions .poly.mult and .poly.quo.rem which make use of compiled code,[3] but could be implemented less efficiently in S as

```
.poly.mult <- function(e1, e2) {
  m <- outer(e1, e2)
  as.vector(tapply(m, row(m) + col(m), sum))
}

.poly.quo.rem <- function(e1, e2) {
  j1 <- length(e1); j2 <- length(e2)
  if(j1 < j2) return(list(quotient = 0, remainder = e1))
  if (j2 == 1) return(list(quotient = e1/e2, remainder = 0))
  e2c <- e2[j2]
  j2 <- j2 - 1
  j1 <- j1 - j2
  quot <- numeric(j1)
  j2s <- 0:j2
  while(j1 > 0) {
    quot[j1] <- f <- e1[j1 + j2]/e2c
    e1[j1 + j2s] <- e1[j1 + j2s] - e2 * f
    j1 <- j1 - 1
  }
  list(quotient = quot, remainder = e1)
```

---

[3] .poly.mult is just the %+% operator we consider in Chapter 6.

```
        }
```

To avoid name clashes these functions could have been defined within the group generic function itself but as that function will be called very frequently we want to keep it as simple as possible. Instead we have defined ordinary functions and used the 'dot name' convention to minimise any naming problem.

Powering is only allowed for non-negative integer exponents and is somewhat inefficiently implemented. Amongst the logical operators only exact equality and inequality will be supported.

The most useful thing for a `Summary` method to do is to signal that the user has probably made a slip.

```
        Summary.polynomial <- function(x, ...) {
          warning("polynomial coerced to numeric vector")
          x <- as.numeric(x)
          NextMethod(.Generic)
        }
```

Methods for functions in the `Math` group like `round` and `signif` are occasionally needed to manipulate the coefficients, but most others are probably best disallowed.

```
        Math.polynomial <- function(x, digits, ...) {
          x <- as.numeric(x)
          switch(.Generic,
                 round = , signif = polynomial(NextMethod(.Generic)),
                 { warning("polynomial coerced to numeric vector")
                   NextMethod(.Generic)})
        }
```

### Coercion to character and printing

The function `as.character` (like most of the system coercion functions) is generic. In the S engines it does not contain a reference to `UseMethod`, but most (though not all) functions with a reference to `.Internal` are generic functions. The list of such functions is apparently different for the new and old S engines; we can only advise readers to experiment!

The following method goes to some trouble to cast the polynomial in a conventional form and provides a result that may also be parsed into a valid S expression. We first define a 'worker' ancillary function which will be useful for the `print` method as well.

```
        makePolyChar <- function(p, rational = F) {
          p <- as.numeric(p)
          tap <- if(rational) .vfrac(abs(p)) else
                   as.character(zapsmall(abs(p)))
          lp <- length(p) - 1
          names(p) <- 0:lp
          if(any(discard <- match(tap, c("0", "-0"), nomatch = F))) {
```

```
Ops.polynomial <- function(e1, e2) {
  if (missing(e2)) {
    ## Unary operators
    e1 <- unclass(e1)
    return(polynomial(NextMethod(.Generic)))
  }
  e1 <- unclass(e1)
  e2 <- unclass(e2)
  e1.op.e2 <-
    switch(.Generic,
           "+" = , "-" = {
             if((j1 <- length(e1)) < (j2 <- length(e2)))
               e1 <- c(e1, rep(0, j2 - j1))
             if(j2 < j1)
               e2 <- c(e2, rep(0, j1 - j2))
             NextMethod(.Generic)
           },
           "*" = .poly.mult(e1, e2),
           "/" = , "%/%" = .poly.quo.rem(e1, e2)$quotient,
           "%%" = .poly.quo.rem(e1, e2)$remainder,
           "^" = if (length(e2) > 1 || e2 < 0 || (e2 %% 1) != 0)
           stop("non positive integer power requested")
           else {
             switch(as.character(e2),
                    "0" = polynomial(1),
                    "1" = polynomial(e1),
                    {
                      m <- e1
                      for (i in 2:e2) m <- .poly.mult(m, e1)
                      m
                    })
           },
           "==" = length(e1) == length(e2) && all(e1 == e2),
           "!=" = length(e1) != length(e2) || any(e1 != e2),

           stop(paste("\"", .Generic, "\"",
                      " is undefined for polynomials",
                      sep = "")))
  polynomial(e1.op.e2)
}
```

**Figure 4.1**: Group method for unary and binary operations with polynomials.

```
    p <- p[!discard]
    if(length(p) == 0) return("0")
    tap <- tap[!discard]
  }
  sp <- c("- ", "+ ")[(p > 0) + 1]
  np <- names(p)
  p <- tap
  p[p == "1" & np != "0"] <- ""
  pow <- paste("x", "^", np, sep = "")
  pow[np == "0"] <- ""
  pow[np == "1"] <- "x"
  stars <- rep("*", length(p))
  stars[p == "" | pow == ""] <- ""
  p <- paste(sp, p, stars, pow, sep = "", collapse = " ")
  while((ch <- substring(p, 1, 1)) == "+" || ch == " ")
    p <- substring(p, 2)
  p
}

as.character.polynomial <- function(x)
  makePolyChar(unclass(x))
```

Some special classes of polynomial are easier to appreciate if their coefficients
are given in rational number form such as ' 3/7 ' rather than floating point. This
option is handled by the helper function, `.vfrac`, not discussed here.

Here is a simple example

```
> x <- polynomial()
> p <- (x^2 + 1)^2
> as.character(p)
[1] "1 + 2*x^2 + x^4"
```

Notice that zero coefficients are not displayed and unit coefficients are suppressed.

Printing a polynomial involves suppressing excessive significant digits, casting the polynomial into character form and displaying it in such a way that if more
than one line is needed the breaks occur at natural places.

```
print.polynomial <- function (x, ..., rational = F)
{
  ow <- max(35, unlist(options("width")) - 1)
  p <- makePolyChar(unclass(x), rational = rational)
  m2 <- 0
  np <- nchar(p)
  while (m2 < np) {
    m1 <- m2 + 1
    m2 <- m2 + ow
    if (m2 < np)
      while (m2 > m1 && substring(p,m2,m2) != " ") m2 <- m2 - 1
    cat(substring(p, m1, m2), "\n")
  }
}
```

```
        invisible(x)
    }
```

The ... argument is needed as the generic function is defined as `print(x, ...)`. Note that the return value must be set `invisible` to avoid an infinite recursion.

## Coercion to functions

One of the useful things to provide for polynomial objects is a method for coercing them to S functions so that they may be evaluated. The simplest way to do this would be first to turn the polynomial into a character string, parse it to an S expression and make that the single line body of a function. A function to do this is

```
    poly2func <- function(p) {  ## example code only (S)
      body <- parse(text = as.character(p))[[1]]
      substitute(function(x) .body, list(.body = body))
    }
```

The S function `parse` can be used to convert a character string into an unevaluated expression. The first component of the result is an object of mode `call` that may be used as the body of the single statement function and we use `substitute` to do the insertion in a simple template. The R version is a little different (see page 68)

```
    poly2func <- function(p) {  ## example code only (R)
      f <- function(x) NULL
      body(f) <- parse(text = as.character(p))[[1]]
      f
    }
```

This simple approach has at least two problems. Using `parse` can be very slow. More importantly the resulting function evaluates its argument in power form and from nearly all points of view it would be better to evaluate the nested multiplication form known as the Horner scheme. The actual method, `as.function.polynomial`, (given in scripts for this chapter) constructs the expressions using low-level tools such as coercion and the constructor function, `call`, and writes the function as a sequence of assignments corresponding to the steps of the Horner scheme. It is quite complicated. An example of the code in action is

```
    > x <- polynomial()
    > p <- (x^2 +1)^2
    > p
    1 + 2*x^2 + x^4
    > fp <- as.function(p)
    > fp
    function (x)
    {
```

```
        p <- 0 + x * 1
        p <- 2 + x * p
        p <- 0 + x * p
        1 + x * p
}
> fp(0:4)                        # numeric argument
[1]    1    4   25 100 289
> fp(x - 1)                      # polynomial argument
4 - 8*x + 8*x^2 - 4*x^3 + x^4
```

Notice that this method solves two problems at once: evaluation of polynomials at numeric arguments and substitution of one polynomial into another.

For R there is an entirely different way of constructing this method that uses the notions of closure and lexical scope.

```
Rpoly <- function (p) {
    function(x) { v <- 0; for (a in rev(p)) v <- a + x * v; v }
}
```

This returns a template function with p in its environment. An example is

```
> rp <- Rpoly(p+1)
> rp(0:4)
[1]    2    5   26 101 290
> rp(x - 1)
5 - 8*x + 8*x^2 - 4*x^3 + x^4
```

We can achieve almost the same effect in S by passing p as a default argument.

```
Spoly <- function(p) {
  f <- function(x, p) {
    v <- 0; for (a in rev(p)) v <- a + x * v; v }
  f$p <- p
  f
}
```

We will see more of these approaches in Section 7.4.


## Orthogonal polynomials

An important use of the polynomial class is to manipulate orthogonal polynomials in regression. The orthogonal polynomials may be constructed as vectors of values using the poly function. When this is called with one argument the value it returns has attributes that provide the coefficients for a two-term recurrence relation between the polynomials to be used. (See the on-line help for poly for the precise details.) The following function calculates the polynomials as a list using the recurrence relation. The norm argument specifies whether the polynomials are to be left as monic or normalized so that the resulting vectors have unit length.

```
poly.orth <- function(x, degree = length(unique(x)) - 1,
                            norm = T)
{
  at <- attr(poly(x, degree), "coefs")
  a <- at$alpha
  N <- at$norm2
  x <- polynomial()
  p <- list(polynomial(0), polynomial(1))
  for(j in 1:degree)
    p[[j + 2]] <-
      (x - a[j]) * p[[j + 1]] - N[j + 1]/N[j] * p[[j]]
  p <- p[-1]
  if(norm) {
    sqrtN <- sqrt(N[-1])
    for(j in 1 + 0:degree) p[[j]] <- p[[j]]/sqrtN[j]
  }
  p
}
```

As a simple example consider the first five orthogonal polynomials on the discrete
set 0:5:

```
> poly.orth(0:5, degree = 4, norm = F)
[[1]]:
1
[[2]]:
-2.5 + x
[[3]]:
3.3333 - 5*x + x^2
[[4]]:
-3 + 13.7*x - 7.5*x^2 + x^3
[[5]]:
1.7143 - 28.571*x + 30.714*x^2 - 10*x^3 + x^4
```

## Exercises

4.3.1 Write a method for the `predict` generic function to evaluate a polynomial
at a given argument x. For efficiency do not coerce the polynomial to
a function. Will your method work with polynomial as well as numeric
arguments?

Given a polynomial $p(x)$ it is occasionally necessary to find the coefficients
of another polynomial, $q(x)$ such that $p(x) \equiv q(x - a)$ for some constant
$a$. Write a one-line function to find $q$ given $p$ and $a$.

4.3.2 Write generic functions `integral` and `derivative` with methods for
objects of class `polynomial` to provide a facility for elementary calculus.

4.3.3 The S function `polyroot` may be used to calculate the zeros of a polynomial. Write a method for the existing generic function `solve` that will
solve polynomial equations of the form $a(x) = b(x)$ using `polyroot`.
Make sure it does something sensible in the cases of degree 0 and 1.

4.3.4 Write a `summary` method for polynomials that will provide the zeros, stationary points and points of inflexion of a polynomial. Give the result some special class, say `polySummary`, and write a print method to display it in a neat and informative way.

4.3.5 Write a `plot` method for polynomials. Allow the user to specify the $x$-region of interest, or deduce it from the polynomial itself by requiring it to cover the real parts of all zeros, stationary points and points of inflexion.

4.3.6 The Hermite polynomials are defined recursively by

$$H_0(x) = 1, \quad H_1(x) = x, \quad H_{n+1}(x) = xH_n(x) - nH_{n-1}(x), \; n \geqslant 1$$

Calculate the first 10 Hermite polynomials and plot them on the same display.

4.3.7 For many types of polynomial certain features are known in advance. For example the context might suggest using $t$ as the display variable rather than $x$ or it may be known in advance that presenting the coefficients in rational form will be appropriate. One way to accommodate these variants is to provide arguments in the generic functions for them, but that option becomes effectively unavailable when the generic function is a system function like `as.character` which does not have a ... argument and is certainly undesirable for functions like `print`.

Another strategy (in some ways preferable) is to attach any special information to the object itself as attributes and to write the method functions to expect to find it as part of the object rather than as additional arguments. In this way the object carries with it as much pertinent information as possible, which is consistent with the object-oriented paradigm.

As an open-ended exercise, adapt the polynomial class so that the polynomial objects may carry optional attributes specifying the display variable and a display protocol (rational or floating point) or any other property you think might be useful. You will need to consider carefully how to resolve conflicts, for example when adding two polynomials with different display variables or protocols.

# Chapter 5

# New-style Classes

The new S engine introduced a very different approach to classes, although backwards compatibility is provided for the classes as described in Chapter 4. The current S-PLUS systems rely very heavily on this backwards compatibility; at present new-style classes are used only at a low level and in the time-series manipulation software. The definitive reference for the new-style classes is Chambers (1998).

*Every* object in the new S engine has a class, and only one class. For example, vectors are of class `numeric`, `integer`, `character`, ... and matrices are of class `matrix`.

All objects in a class must have the same structure. This is not true of many old-style classes; for example objects of class `lm` are lists which may or may not have a component named `x`, and the attributes mechanism is often used to append varying amounts of information. This *can* be accommodated in the new-style classes by specifying a class as composed of elements of class `"list"` or `"structure"`, but the advantages of the new-style classes accrue from a tighter specification.

All methods for a new-style generic function must have exactly the same formal arguments. Again, flexibility is possible, via a ... formal argument, but it is harder to achieve.

A substantial amount of information about classes is stored in *metadata*, special S objects stored in a special area of the S database (usually directory `.Data/__Meta`, though programmers need never use this directory directly).

Having drawn to the reader's attention that new-style classes really are different and have to be used in a less casual way, we should immediately point out that they do have corresponding advantages, although these may take some time for the beginning programmer to appreciate fully. The author of the new engine, Dr John Chambers, has kindly given us a tentative list of those that he sees in rough order of importance, namely

1. The new class mechanism has greater uniformity than the old and the new engine has many more tools to make programming easier and clearer,

2. Methods can be defined for data types (for example, vectors) that were not classes in the old engine (although the function `data.class` went some way to affording this),

3. The metadata information allows the code to access and use the language explicitly, for example finding methods and class properties directly,

4. Virtual classes allow methods to be written for distinct actual classes that share some common abstract property, for example that they allow componentwise indexing.

To this we might add a few more immediately appreciated qualities that first occurred to us. By having a fixed structure fewer calculations are needed at run-time, and it is possible to check the validity of objects against the class structure and so avoid storing incorrectly-structured objects.

New-style classes are made up of *slots*. These are similar to but distinct from components of a list, in that the number of slots and their names and classes are specified when a class is created: objects are extracted from slots by the @ operator. Exact matching of slot names is used, unlike $ for list components.

New-style classes are optional: old-style classes still exist and are still widely used. We hope though that the new mechanisms *will* be tried for new projects.

## 5.1   Creating a class

A class is created by the function `setClass`. Its first argument is the name of the class, and its `representation` argument specifies the slots. For example, a class to represent the spatial locations of fungi in a field might have

```
setClass("fungi", representation(x = "numeric", y = "numeric",
                                 species = "character"))
```

Note how the class is made up by combining other classes. Sometimes we might wish to do this recursively, so we might instead have

```
setClass("xyloc", representation(x = "numeric", y = "numeric"))
setClass("fungi", representation("xyloc", species = "character"))
```

This gives the same structure, as giving an unnamed argument to the function `representation` requests that the slots from that class be included.

Once a class has been created, it can be examined by `getClass`.

```
> getClass("fungi")

Slots:
          x          y      species
  "numeric"  "numeric"  "character"

Extends:
Class "xyloc" by direct inclusion
```

The last statement means that wherever an object of class "xyloc" is needed, one of class "fungi" can be used and the corresponding slots will be extracted and used.

To create an object from the class, use new, for example

```
field1 <- new("fungi", x = runif(25), y = runif(25),
              species = sample(letters[1:5], 25, rep=T))
```

We need to specify the contents of each slot unless we have set up a suitable prototype (see later).

To examine the new object, just type its name:

```
> field1
An object of class "fungi"

Slot "x":
 [1] 0.960659 0.937460 0.044102 0.764619 0.705858 0.503551
 [7] 0.928648 0.840273 0.547102 0.487805 0.398985 0.263520
[13] 0.925925 0.428515 0.960061 0.298338 0.577216 0.488645
[19] 0.159737 0.182527 0.213183 0.265986 0.732724 0.643868
[25] 0.897490

Slot "y":
 [1] 0.499815 0.576944 0.905814 0.014425 0.746566 0.421574
 [7] 0.494986 0.725713 0.071926 0.279652 0.975037 0.509670
[13] 0.772228 0.980534 0.714942 0.471561 0.119848 0.716741
[19] 0.303823 0.775617 0.136685 0.017758 0.791863 0.723880
[25] 0.880876

Slot "species":
 [1] "b" "a" "d" "c" "b" "c" "d" "a" "a" "d" "a" "c" "b" "b"
[15] "c" "d" "d" "d" "a" "d" "c" "a" "d" "d" "a"
```

This calls show (not print) to display the object. We will write a better display function for "fungi" objects later.

Users will rarely see a call to new, as it is usually used only within functions that create objects. We might for example use

```
fungi <- function(x, y, species)
    new("fungi", x = x, y = y, species = species)
```

It is tempting to check here that the arguments are sensible, for example of the same length. However, that is a sanity check we should apply to all objects of the class, covered in the next subsection.

A class can be removed by a call to removeClass.

## Validity checking

Our specification of the class "fungi" is incomplete: we want all the slots to be
of the same length. We could ensure this by either of

```
setClass("fungi", representation("xyloc", species = "character"),
                   validity = validFungi)
setValidity("fungi", validFungi)
```

where the checking function is

```
validFungi <- function(object)
{
  len <- length(object@x)
  if(length(object@y) != len || length(object@species) != len)
    return("mismatch in lengths of slots")
  else return(TRUE)
}
```

Care is needed here: this sets the checking function to a copy of the *current*
version of the function named. If you change the function, use setValidity
again.

Objects are checked for validity whenever they are assigned on a permanent
database.

## Looking at classes

We have already seen that getClass will give the current definition, and this
includes a validity function if one is defined. A more complete picture can be
obtained by dumpClass, which writes all the information to a file:

```
> dumpClass("fungi")
[1] "fungi.class.q"
> !cat fungi.class.q  # under UNIX, at least
setClass("
fungi
",
  representation=
representation("xyloc", species = "character")
,
    validity = function(object)
{
        len <- length(object@x)
        if(length(object@y) != len ||
            length(object@species) != len)
                return("mismatch in lengths of slots")
        else return(TRUE)
}
)
```

This gives all the information needed to re-create the class.

Information on the slots can be obtained by

```
> getSlots("fungi")
        x          y       species
  "numeric"  "numeric"  "character"
> slotNames("fungi")
[1] "x"        "y"          "species"
> hasSlot(field1, "x")
[1] T
```

The first two of these can also be applied to an object of the class.


## Prototypes

Each class has a prototype stored with its definition; the default prototype is to take for each slot a new object of the appropriate class. This is given by `new` if no other arguments are supplied:

```
> new("fungi")
An object of class "fungi"
Slot "x":
numeric(0)
Slot "y":
numeric(0)
Slot "species":
character(0)
```

Sometimes another prototype would be more appropriate, and this can be specified by the `prototype` argument of `setClass` or at a later stage by a call to `setClassPrototype`.

A class can have a prototype but not a representation. One such class is `"sequence"`,

```
> getClass("sequence")

No slots; prototype of class "integer"
> new("sequence")
An object of class "sequence"

numeric(0)
```

This is a class like `"integer"` but with no implied promise that it could be used where an integer vector was required.

For classes without slots the function `unclass` changes the class of its argument to that of the prototype. Thus for an object of class `"sequence"`, `unclass` returns an object of class `"integer"`.

## Virtual classes

Virtual classes usually have neither representation nor prototype, and normally there are no objects of that class. They sound useless, but are made useful by the inheritance between classes. Consider the class named for vectors with names.

```
> getClass("named")

Slots:
     .Data      .Names
  "vector" "character"

Extends:
Class "structure" by direct inclusion
Class "list" by explicit test, coerce
Class "vector" indirectly through class "structure"
Class "logical" indirectly through class "vector"
Class "single" indirectly through class "vector"
Class "integer" indirectly through class "vector"
Class "numeric" indirectly through class "vector"
Class "complex" indirectly through class "vector"
Class "string" indirectly through class "vector"
Class "character" indirectly through class "vector"
    ....
```

This is able to extend all the basic vector classes by including a slot of class "vector", which is also a virtual class.

```
> getClass("vector")
Virtual Class

No slots; prototype of class "NULL"

Extends:
Class "complex" as a possible instance of the virtual class
Class "string" as a possible instance of the virtual class
Class "character" as a possible instance of the virtual class
Class "numeric" as a possible instance of the virtual class
Class "logical" as a possible instance of the virtual class
Class "list" as a possible instance of the virtual class
    ....
```

Having "vector" as a virtual class allows methods to be written that just assume that a class has a sequential index of 'elements'. (This abstraction can make coding very economical.) Note that lists are a class extending "vector": they are another virtual class.

Creating virtual classes is easy: either give no representation or include the special class "VIRTUAL" in the representation.

One exception to the rule that virtual classes do not have any direct members is in fact a very common one: old-style classes are effectively implemented as virtual classes, for example

```
> getClass("lm")
Virtual Class

No slots; prototype of class "NULL"
```

but see Section 5.4 for further details.

## 5.2   Inheritance

The power of a class-oriented language stems from the relationships between classes. Note that in Chapter 4 inheritance applied to objects that could belong to multiple classes. Here objects have precisely one class, and inheritance applies to classes.

In S there are 'is' and 'as' relationships. Class A 'is' (extends, inherits from) class B if is(x, "B") is true for all objects x of class A; in essence this means that whenever an object of class B is required, any x from A is acceptable. This is reflected in

```
> extends("fungi", "xyloc")
[1] T
```

and so for all members x of class "fungi", is(x, "xyloc") is true.

This relationship can also be conditional: only some objects of class "vector" are of class "integer", for example

```
> c(is(10, "integer"), is(10.5, "integer"))
  [1] T F
```

However extends reports

```
> extends("vector", "integer")
[1] T
```

How can this be? In fact extends is being generous, and adding an argument

```
> extends("vector", "integer", maybe = NA)
[1] NA
```

shows a more complete picture. Of course in the other direction inclusion is always true:

```
> extends("integer", "vector", maybe = NA)
[1] T
```

S gets its information on 'is' relationships from the class definitions, specifically the inclusion of one class in the definition of another, and from setIs relationships. We can use[1] setIs to set conditional inheritance, for example

---

[1] you will need to 'own' the classes to do this; as these classes are in database "main" this will fail if tried.

```
setIs("vector", "integer",
      test = function(object) class(object)=="integer")
```

Inheritance is transitive, and most of the relationships listed by `getClass` come through transitivity. The list given by `getClass` is in order of 'closeness'[2] to the given class.

We have seen that class `"integer"` extends class `"numeric"`. This relies on some coercion, as the internal representation of the classes is different. If necessary such coercion can be prescribed by the `coerce` argument to `setIs`.

Sometimes transitivity gives inheritance that we do not want, or conditional inheritance via a series of tests which we know cannot all be passed. We can deny inheritance by using `test = F` in a call to `setIs`.

### Coercion

No objects of class `"numeric"` are regarded as being of class `"integer"`

```
> extends("numeric", "integer", maybe = NA)
[1] F
```

as the representations are different and implicit coercion might lose information. What we need is a way to *coerce* objects of class `"numeric"` to class `"integer"`, and this is provided by the function `as`

```
> as(c(1., 2., 3., 4.4), "integer")
[1] 1 2 3 4
```

There is a third argument to `as`, `coerce`, which defaults to true and ensures coercion, otherwise `as(x, class, F)` will succeed only if `is(x, class)` is true (and that might involve some coercion). Thus if we try

```
x <- 3; class(x); as(x, "numeric", F)
x <- as.double(3); class(x); as(x, "integer", F)
```

the first is of class `"integer"` but will succeed by coercing the internal representation, whereas the second will fail.

By default `as` calls the generic function `coerce`, and that derives its information from `setAs` declarations.

## 5.3  Generic and method functions

We wanted a better display for objects of class `"fungi"`. Automatic printing is done by `show`, a function with a single argument, `object`. We can write a suitable method function and declare it by

---

[2]there is some explanation of this on pages 319–320 of Chambers (1998).

```
functionArgNames("show")
[1] "object"
show.fungi <- function(object) {
  tmp <- rbind(x = format(round(object@x, 2)),
               y = format(round(object@y, 2)),
               species = object@species)
  dimnames(tmp)[[2]] <- rep("", length(object@x))
  print(tmp, quote=F)
  invisible(object)
}
setMethod("show", "fungi", show.fungi)
```

Notice that the method no longer needs to be called show.fungi (although that is a convenient reminder of its purpose) but that it does have to be declared as a method by a call to setMethod.

This declares the current version of show.fungi as the method: changes to show.fungi would not be reflected in the method used. In fact after the call to setMethod has been made the function show.fungi could be discarded as a copy of it exists in the metadata.

Further thought suggests that this strategy is not the best as we may also need a method for print (for example to cater for users who invoke print explicitly in functions or loops), which is an old-style generic function. So we may prefer to use

```
print.fungi <- function(x, digits=2) {
  tmp <- rbind(x = format(round(x@x, digits)),
               y = format(round(x@y, digits)),
               species = x@species)
  if(!is.null(tmp)) {
    dimnames(tmp)[[2]] = rep("", length(x@x))
    print(tmp, quote=F)
  } else cat("empty object of class fungi\n")
  invisible(x)
}
setMethod("show", "fungi", function(object) print.fungi(object))
```

Then we can say field1 or show(field1) or print(field1) to equal effect. The advantage of print is that it can be more flexible by not being bound to a single argument. The advantage of show is that we do not need to print at all: we could draw a picture.

We would also like a plot method for this class. However, plot is a confusing mixture of the old and new styles:

```
> functionArgNames("plot")
[1] "x"   "y"   "..."
> plot
function(x, ...) UseMethod("plot")
> selectMethod("plot", "fungi")
function(x, y, ...) {
```

```
xyCall(x, y, function(x, y, xlab, ylab, ...)
    .Internal(plot("zplot", x = x, y = y, xlab = xlab,
                   ylab = ylab, ...), "call_S_Version2"),
                xexpr=substitute(x), yexpr=substitute(y), ...)
}
```

`selectMethod` correctly says what will be called. We need to write and declare a method by, for instance

```
plot.fungi <- function(x, y, ...) {
    oldpar <- par(pty="s")
    on.exit(par(oldpar))
    plot(x@x, x@y, type="n", xlab="x", ylab="y")
    text(x@x, x@y, labels=x@species)
    title(deparse(substitute(x)))
}
setMethod("plot", "fungi", plot.fungi) ## not recommended
```

Note that we do need exactly this set of arguments, even if we do not use `y`. Trying to use a different set of arguments (even different names) will give warnings and sometimes unpredictable behaviour: for example `match.call` is unlikely to work. Since we only specified the class for one argument of `plot`, our method will work for any (including missing) argument `y`. Perhaps a safer approach is

```
setMethod("plot", signature(x="fungi", y="missing"), plot.fungi)
```

that insists that `y` really is missing.

Methods for replacement functions are set by `setReplaceMethod` rather than `setMethod`. We give an example on page 117.

## Method dispatch

With the old-style classes, methods are selected[3] on the principal (usually the first) argument. In the new engine, two or more arguments can be used to select the method. The second argument to `setMethod` is a *signature* specifying the classes of the named or unnamed arguments to be used in method dispatch. Giving an explicit class is a shortcut to specify the class of the first argument, but otherwise the signature should be constructed via a call to `signature`.

The function `selectMethod` shows the method that would be selected for a particular signature, and `showMethods` will list all the methods for the function (which can be many). Function `hasMethod` will say if a non-default method will be found for that signature, and `existsMethod`, `getMethod` (or its variant `showMethod`) and `findMethod` look for a method which is defined for the specified signature (rather than being found by inheritance). Function `dumpMethod` will write the selected method for the given signature to a file as a call to `setMethod`.

A method can be removed by a call to `removeMethod`.

---

[3] with a few exceptions, see page 78.

We can now find the method of `coerce` that `as` uses to coerce numeric vectors to class `"integer"`.

```
> selectMethod("coerce", signature("numeric", "integer"))
function(object, to)
as.integer(object)
```

The details of how the method shown by `selectMethod` is selected are given in Chambers (1998, §4.9.6). If a method is found that has signature exactly matching the supplied classes, it is used. Otherwise the method with the 'closest' signature is used. For a single class, the ordering of other classes in closeness is the list given by calling `getClass`. Thus when a method is selected on the basis of a single argument, S looks for a method in turn for each of the classes that extend the supplied class. As a default method has signature `"ANY"`, it will be found at the end of the match list. Where two or more arguments are involved in a signature, the search uses lexicographical ordering.

The same selection process is used for method dispatch.

### Generic functions

Normally there is no need to declare a generic function: setting a method for a function automatically makes it generic with the old definition as the default method and the arguments of the old method defining the (fixed) set of arguments of the generic function. The function `isGeneric` will reveal if this has happened.

It is also possible to declare a generic function directly via `setGeneric`. An example is

```
setGeneric("lda", function(x, y, ...) {
        res <- standardGeneric("lda")
        res@call = match.call()
        res
    })
```

which (unlike `UseMethod`) allows some post-processing of the calls to each method.

## 5.4   Old-style classes

We have seen that old-style classes are effectively virtual classes in the new style. The story is actually a little more subtle. They appear as virtual classes for the reason that, like virtual classes, they lack an explicit implementation.

We may check for the existence of an old-style class using

```
> isOldClass("glm")
[1] T
```

Set old-style classes via a call to the replacement function `oldClass<-` rather than by calling `class` or setting attribute `"class"`, perhaps by returning a `structure`. (This used to be a favourite coding idiom of one of us and made conversion to the new engine somewhat tedious.) Thus `lm.fit.qr` now ends

```
    oldClass(fit) <- cl
    fit
}
```

Use `oldUnclass` rather than `unclass` for old-style classes. As all S objects must have a class, this does not remove the class but it does remove the attribute marking the old-style class.

Objects can now have only one old-style class, whereas `glm` objects, for example, used to have class `c("glm", "lm")`. This can be emulated by setting explicit inheritance, for example by

```
    setOldClass(c("glm", "lm"))
```

but that will apply to all objects of the class (which it probably did anyway).

## 5.5   An extended statistical example revisited

Section 4.2 discusses how we converted the function `lda` from a single function to a generic for several different types of argument, and added suitable method functions. Here we discuss how to take that work and to convert it to the new-style class structure of the new S engine.

Some fundamental design (in our case re-design) decisions have to be made early, for new-style classes are rigorously defined, and all methods for a new-style generic must have identical call sequences. First the class. Objects of the old class `"lda"` are lists with components

```
    prior, counts, means, scaling, lev, svd, N, call
```

and perhaps `terms` (only present in objects created by `lda.formula`). We can convert this to a new-style class by giving the objects the following slots

```
    setClass("lda", representation(prior =    "named",
            counts =  "named",    means =    "matrix",
            scaling = "matrix",   lev =      "character",
            svd =     "numeric",  N =        "integer",
            call =    "call") )
```

Note that we have to specify the classes of the slots very carefully: numeric and integer slots do not have names, but slots of class `"named"` do. We do not include a slot for `terms`, as `"terms"` is an old-style class with no fixed representation. (We could have used `"list"` for it).

Our next decision is the set of arguments for all the methods. We had some choice here, and experimented with naming all the possible arguments for each method or making use of ' ... '. We settled on

```
lda <- function(x, y, ...)
    stop("lda not implemented for class ", class(x))
```

as a function that will become the default method. The idea is that x and y will represent either the matrix x and grouping or formula and data, and all other arguments must be named exactly. Since the previous lda.default is not going to be one of the methods, we rename it to have call

```
lda1 <- function(x, grouping, prior = NULL, tol = 1.0e-4,
                 method = c("moment", "mle", "mve", "t"), CV=F,
                 nu = 5, ...)
{
    ....
}
```

We can now begin to set up methods. Every object in the new S engine has a class, which eases the process. The methods for matrices and data frames are easy:

```
lda.data.frame <- function(x, y, ...)
{
    x <- as.matrix(x)
    callGeneric()
}

lda.matrix <- function(x, y, ...)
{
    x <- as(x, "matrix")
    dots <- list(...)
    if(hasArg(subset)) {
        subset <- dots$subset
        x <- x[subset, , drop = F]
        y <- y[subset]
    }
    if(hasArg(na.action)) {
        na.action <- dots$na.action
        dfr <- na.action(data.frame(g = y, x = x))
        y <- dfr$g
        x <- dfr$x
    }
    res <- lda1(x = x, grouping = y, ...)
    res@call <- match.call()
    res
}

setMethod("lda", "matrix", lda.matrix)
setMethod("lda", "data.frame", lda.data.frame)
```

Note that rather than call the function NextMethod, we call the generic again with callGeneric. The generic is no longer the function lda that we defined, which has become the default method (and listing functions can be misleading).

The other features to note are that `call` is now a slot set by the `@` operator, and the way missing arguments are handled. Missing arguments are definitely treated differently: explicit arguments of generic functions which are missing are of class `"missing"` and `missing()` does not work for them.

Arguments in `...` can be checked for by `hasArg` provided they are named in the call. That was our design decision for handling `subset` and `na.action` with old-style classes (when they followed `...`) so is safe here. Note that these two arguments (if present) are passed down to `lda1` and swallowed by its unused `...` argument.

We chose to make more extensive changes for a `formula` method.

```
lda.formula <- function(x, y, ...)
{
  data <- as.data.frame(y)
  m <- match.call(expand.dots = F)
  m$... <- NULL
  dots <- list(...)
  names(m)[2:3] <- c("formula", "data")
  if(hasArg(subset)) m$subset <- dots$subset
  if(hasArg(na.action)) m$na.action <- dots$na.action
  m[[1]] <- as.name("model.frame")
  m <- eval(m, sys.parent())
  Terms <- attr(m, "terms")
  y <- model.extract(m, "response")
  x <- model.matrix(Terms, m)
  xint <- match("(Intercept)", dimnames(x)[[2]], nomatch = 0)
  if(xint > 0) x <- x[, -xint, drop = F]
  res <- lda1(x <- x, grouping = y, ...)
  Call <- match.call()
  Call$x <- as.call(attr(Terms, "formula"))
  res@call <- Call
  res
}
setMethod("lda", "formula", lda.formula)
```

First, the argument names have changed (they have to be the same as the other methods) and so the lazy way of calling `model.frame` used on page 85 has to be replaced by an explicitly constructed call. Second, we cannot add the `terms` component, as it is not in the design of the class object. So we note that we will have to cope with this in methods for the new class. In fact there was a lazier way to do this, by not calling `lda.formula` directly.

```
setMethod("lda", "formula",
          function (x, y, ...)  oldlda.formula(x, y, ...))

oldlda.formula <- function(formula, data = NULL, ...,
                           subset, na.action = na.fail)
{
    ....
```

```
        res <- lda1(x, grouping, ...)
        Call <- match.call()
        Call$x <- as.call(attr(Terms, "formula"))
        Call[[1]] <- as.name("lda")
        res@call <- Call
        res
    }
```

The snag with this approach is that the call needs to be manipulated. Note that we manipulate it anyway, to expand out any `.` on the right-hand side of the formula for future use.

The changes to `lda1` are mainly in constructing the returned object. We have also to ensure that `counts` has the right class ( `table` returns a one-dimensional array).

```
    lda1 <- function(x, grouping, prior = NULL, tol = 1.0e-4,
                     method = c("moment", "mle", "mve", "t"), CV=F,
                     nu = 5, ...)
    {
        ....
        lev <- lev1 <- levels(g)
        counts <- as.vector(table(g))
        ....
        names(prior) <- names(counts) <- lev1
        ....
        res <- new("lda")
        res@prior <- prior
        res@counts <- counts
        res@means <- group.means
        res@scaling <- scaling
        res@lev <- lev
        res@svd <- X.s$d[1:rank]
        res@N <- n
        res
    }
```

We need some changes in the methods, apart from replacing `$` by `@` to reflect the change from a list to a representation with slots.

We need to declare a means of displaying the object, as `show` will not automatically call `print.lda` but rather would list the slots.

```
    print.lda <- function(x, ...)
    {
        if(!is.null(cl <- x@call)) {
            names(cl)[2:3] <- ""
            cat("Call:\n")
            dput(cl)
        }
        cat("\nPrior probabilities of groups:\n")
        print(x@prior, ...)
```

```
    cat("\nGroup means:\n")
    print(x@means, ...)
    cat("\nCoefficients of linear discriminants:\n")
    print(x@scaling, ...)
    svd <- x@svd
    names(svd) <- dimnames(x@scaling)[[2]]
    if(length(svd) > 1) {
      cat("\nProportion of trace:\n")
      print(round(svd^2/sum(svd^2), 4), ...)
    }
    invisible(x)
}


setMethod("show", "lda", function(object) print.lda(object))
```

Next, we need to recover from having no `terms` component. We used the
following scheme in `predict.lda`.

```
predict.lda <-
function(object, newdata, prior = object@prior, dimen,
         method = c("plug-in", "predictive", "debiased"), ...)
{
  if(!inherits(object, "lda")) stop("object not of class lda")
  if((missing(newdata) || !is(newdata, "model.matrix")) &&
      is.form(form <- object@call[[2]])) { #
    # formula fit
    if(missing(newdata)) newdata <- model.frame(object)
    else newdata <- model.frame(delete.response(terms(form)),
                                newdata,
                                na.action = function(x) x)
    x <- model.matrix(delete.response(terms(form)), newdata)
    xint <- match("(Intercept)", dimnames(x)[[2]], nomatch = 0)
    if(xint > 0) x <- x[, -xint, drop=F]
  } else { #
    # matrix or data-frame fit
    if(missing(newdata)) {
      if(!is.null(sub <- object@call$subset))
        newdata <- eval(parse(text=paste(deparse(object@call$x),
                       "[", deparse(sub),",]")), sys.parent())
      else newdata <- eval(object@call$x, sys.parent())
      if(!is.null(nas <- object@call$na.action))
        newdata <- eval(call(nas, newdata))
    }
    if(is.null(dim(newdata)))
      dim(newdata) <- c(1, length(newdata))   # a row vector
    x <- as.matrix(newdata)                # to cope with data frames
  }
    ....
is.form <- function(x) is.call(x) && (x[[1]] == "~")
```

We test for an object produced by `lda.formula` by testing the first argument of the call (which is of class `"call"`, not `"formula"` as one might expect). We can re-create the `terms` object from the formula. Note that we do need a method for `model.frame`, as we no longer have a list object, and the names of our arguments in the call are non-standard for a model-fitting function.

```
model.frame.lda <-
function(formula, data = NULL, na.action = NULL, ...)
{
  oc <- formula@call
  oc[[1]] <- as.name("model.frame")
  names(oc)[2:3] <- c("formula", "data")
  oc$prior <- oc$tol <- oc$method <- oc$CV <- oc$nu <- NULL
  oc[[1]] <- as.name("model.frame")
  if(length(data)) {
    oc$data <- substitute(data)
    eval(oc, sys.parent())
  }
  else eval(oc, list())
}
```

Finally, `update` will no longer work, and we need

```
formula.lda <- function(object) object@call[[2]]
```

and `update.lda` which is `update.default` with `object$call` replaced by `object@call`. One problem with using new-style classes when the S model-fitting functions have not been converted is that many of the utility functions expect a list as the class object. Perhaps in due course they will be updated to look for slots or lists.

## 5.6   Group methods and another polynomial class

The new S engine has similar group methods to those we explored in Section 4.3. There are some small differences, and the current groupings are shown in Table 5.1.

The members of the groups can be found by calls to `getGroupMembers`. Note that a group (such as `Ops`) can be defined in terms of other groups. One reason for having more groups is the need to have exactly the same set of arguments: in the new S engine `log` has only one argument, and `round` and `signif` have been moved out of group `"Math"` to a group of functions of two arguments. As before, `"-"` is both a unary and binary operator; if a member is used as a unary operator the missing argument, the second, has class `"missing"`.

Methods may be set for whole groups by `setMethod` just as for single generic functions. However, to define a new generic for a whole group, use `setGroupGeneric`, and use `setGroup` to insert a function into a group.

| Group | Component functions |
|---|---|
| Ops | `Arith, Compare, Logic` |
| Arith | `+, -, *, /, ^, %/%, %%` |
| Compare | `<, <=, >, >=, ==, !=, compare` |
| Logic | `&, |, !` |
| Math | `abs, acos, acosh, asin, asinh, atan, atanh, ceiling, cos, cosh, cumsum, exp, floor, gamma, lgamma, log, sin, sinh, tan, tanh, trunc` |
| Math2 | `round, signif` |
| Summary | `all, any, max, min, prod, range, sum` |

**Table 5.1**: The group generic functions and their components.

### Polynomial objects

Let us consider a new-style class for polynomials similar to that considered in Section 4.3. Only some of the code is shown here, but the full version is in the on-line scripts.

In this version the object will carry information not only about its coefficients but also on how it is to be represented in character form (with coefficients as rational or floating-point numbers). It is important that the leading coefficient be non-zero (unless it is the zero polynomial) so in writing a constructor function for polynomial objects we will use a helper function, `.clip`, to remove any leading zero coefficients. We also use a slot to say if the coefficients are rational, and give a validity function.

```
setClass("polynomial",
         representation(coef="numeric", rat="logical"),
         validity = function(object) {
           coef <- object@coef; rat <- object@rat
           if(length(coef) > 1 &&
              abs(coef[length(coef)]) < .Machine$double.eps)
             return("zero leading coefficient")
           else if(length(rat) > 1 || !is(rat, "logical"))
             return("improper rationality marker")
           else return(TRUE)
         })

.clip <- function(p) {
  p <- as(p, "numeric")
  if((j <- length(p)) == 0) return(0)
  while(j > 1 && p[j] == 0) j <- j - 1
```

```
   p[1:j]
}

polynomial <- function(coef = c(0, 1), rat = all(coef %% 1 == 0))
   new("polynomial", coef = .clip(coef), rat = rat)

setAs("polynomial", "numeric", function(object) object@coef)
```

Note that class `"polynomial"` does not inherit from class `"numeric"` (although it would do so if the `coef` slot were unnamed), so we give a method to coerce to `"numeric"`.

Indexing operations on polynomial objects are enabled using

```
setMethod("[", "polynomial", function(x, ..., drop = T)
          polynomial(x@coef[..., drop = drop], rat = x@rat))

setReplaceMethod("[", signature(x = "polynomial"),
                 function(x, ..., value) {
                   cx <- as(x, "numeric")
                   cx[...] <- as(value, "numeric")
                   polynomial(cx, rat = x@rat)
                 })
```

Notice how we use calls to the constructor function, `polynomial`, mainly to ensure that the leading coefficient always remains non-zero.

For the `Math` and `Math2` groups our policy remains to allow those which might be occasionally needed and to disallow (with a message) those which will only be used under some misapprehension by the user.

```
setMethod("Math", "polynomial",
    function(x) {
      switch(.Generic,
             ceiling =, floor =,
             trunc = polynomial(callGeneric(x@coef), rat = x@rat),
             stop(paste(.Generic, "not allowed on polynomials"))
             )}
       )

setMethod("Math2", "polynomial",
          function(x, digits)
          polynomial(callGeneric(x@coef, digits), rat = x@rat)
          )
```

In defining the operations of arithmetic it is convenient to begin with a helper function that does most of the decision-making.

```
.ArithPolynomial <- function(e1, e2) {
  rat <-
    (if(is(e1, "polynomial")) e1@rat else all(e1 %% 1 == 0)) &&
    (if(is(e2, "polynomial")) e2@rat else all(e2 %% 1 == 0))
```

```
e1 <- as(e1, "numeric")
e2 <- as(e2, "numeric")
e1.op.e2 <- switch(.Generic,
  "+" = ,
  "-" = {
    if((j <- length(e1) - length(e2)) < 0)
      e1 <- c(e1, rep(0, -j))
    else if(j > 0) e2 <- c(e2, rep(0, j))
    callGeneric(e1, e2)
  },
  "*" = .poly.mult(e1, e2),
  "/" = ,
  "%/%" = .poly.quo.rem(e1, e2)$quotient,
  "%%" = .poly.quo.rem(e1, e2)$remainder,
  "^" = if(length(e2) > 1 || e2 < 0 || (e2 %% 1) != 0) {
    stop("positive integer powers only")
  } else {
    switch(as.character(e2), # needed as we want a default
           "0" = 1, "1" = e1,
           { m <- e1; for(i in 2:e2) m <- .poly.mult(m, e1)
             m
           })
  })
  polynomial(e1.op.e2, rat = rat)
}
```

If this function had been left anonymous and used directly in the method definition
it would not occupy the working namespace, but as we will need to use it several
times we will define it as a free-standing function but as usual give it a "dot" name
to indicate it is not intended to be used directly.

Arithmetical operations are now relatively straightforward; the unary operations can be handled by a very simple method function and in all other cases
when either operator has class "polynomial", polynomial operations will be
used. This allows us to avoid specifying separate methods for the special cases
where the second operator has any one of a variety of numerical classes. It does,
however, imply that if a numerical vector is combined with a polynomial in arithmetical operations the vector is silently promoted to polynomial status.

```
setMethod("Arith", signature(e1 = "polynomial", e2 = "missing"),
          function(e1, e2) {
            e1 <- unclass(e1)
            polynomial(callGeneric())
          })

setMethod("Arith", signature(e1 = "polynomial"),
          .ArithPolynomial)
setMethod("Arith", signature(e2 = "polynomial"),
          .ArithPolynomial)
```

Only exact equality and inequality are allowed between polynomials.

```
.ComparePolynomial <- function(e1, e2) {
  e1 <- as(e1, "numeric")
  e2 <- as(e2, "numeric")
  switch(.Generic,
          "==" = length(e1) == length(e2) && all(e1 == e2),
          "!=" = length(e1) != length(e2) || any(e1 != e2),
          stop("unsupported comparison of polynomials"))
}

setMethod("Compare", signature(e1 = "polynomial"),
          .ComparePolynomial)
setMethod("Compare", signature(e2 = "polynomial"),
          .ComparePolynomial)

setMethod("Logic", signature(e1 = "polynomial"), function(e1, e2)
          stop("logical operations not allowed on polynomials"))
setMethod("Logic", signature(e2 = "polynomial"), function(e1, e2)
          stop("logical operations not allowed on polynomials"))
```

A very natural operation to perform on polynomials of one variable is to plot them. The `plot` method we supply for the polynomial class is an old-style method. However, methods for other functions such as `lines`, `points`, `text` and a few others[4] may be covered by a method for the function `xyCall`.

```
setMethod("xyCall", signature("polynomial", "missing"),
          function(x, y, FUN, ..., xexpr, yexpr) {
              pu <- par("usr")
              x0 <- seq(pu[1], pu[2], len = 300)
              y0 <- predict(x, x0)
              y0[y0 <= pu[3] | y0 > pu[4]] <- NA
              FUN(x0, y0, ...)
          })
```

(Note that the function's third argument is `FUN` and not `f` as appears in Chambers 1998, p. 335.) In some ways this generic function is like a group generic. The arguments `xexpr` and `yexpr` are passed through to the method as the actual argument expressions used for the `x` and `y` arguments and might be useful, for example, for labelling.

Coercion of polynomials to character does not involve any different ideas or techniques from those needed for the old engine, but coercion to function opens up a new and initially surprising possibility. We use a `setIs` declaration to establish inheritance of polynomial objects from functions. Essentially it provides a way of constructing a function object which will be used whenever a polynomial object is used in a function context.

---

[4]No complete official list has been issued. John Chambers has informed us that it includes `approx`, `chull`, `labclust`, `lines`, `lowess`, `plot`, `points`, `polygon`, `spline`, `text` and `xysort`.

```
setIs("polynomial", "function", coerce =
    function (object) {
        p <- as.name("p")
        x <- as.name("x")
        object <- as(object, "numeric")
        if ((an <- length(object)) == 1) {
            object <- c(object, 0)
            an <- 2
        }
        statement <- expression()
        statement[[i <- 1]] =
            structure(list(p, call("+", object[an-1],
                call("*", x, object[an]))), mode = "<-")
        for (ai in rev(object)[-(1:2)])
            statement[[i <- i + 1]] <-
                structure(list(p, call("+", ai, call("*", x, p))),
                           mode = "<-")
        statement[[i]] <- statement[[i]][[2]]
        statement <-
            if(an == 2) statement[[1]]
            else as(statement, "{")
        substitute(function(x) body, list(body = statement))
    })
```

*Some simple polynomial examples*

The best way to understand what software such as the polynomial library can do is to try it out on simple examples. We invite the reader to do this but give a glimpse here.

```
> x <- polynomial()
> p <- (x - 1)^3
> p
- 1 + 3*x - 3*x^2 + x^3
```

We could coerce this polynomial to function explicitly:

```
> as(p, "function")
function(x)
{
        p <- -3. + x * 1.
        p <- 3. + x * p
        -1. + x * p
}
```

This coercion has been established by the `setIs` declaration. If the function object were not needed explicitly, it could be generated implicitly by using the polynomial object as if it were a function (which in a sense it now is). Moreover the statements inside the function above are interpretable for polynomial as well as numeric arguments, so in passing we have solved the problem of substituting one polynomial into another. The following illustrates this idea.

```
> p(-3:4)                    # numerical argument
[1] -64 -27  -8  -1   0   1   8  27

> p(x+1)                     # polynomial argument: translation
x^3

> p(p)                       # functional iteration.
 - 8 + 36*x  - 90*x^2 + 147*x^3 - 171*x^4 + 144*x^5 - 87*x^6
 + 36*x^7 - 9*x^8 + x^9
```

There is automatic use of rational expressions when all the coefficients are rational.

```
> p/3
 - 1/3 + x - x^2 + 1/3*x^3
```

The simple heuristic adopted is that the result of an operation on polynomials is represented as rational if all polynomials are already rational and any numeric operators are whole numbers:

```
> p/3 + 1.1
0.7666667 + x - x^2 + 0.3333333*x^3
```

(Of course the user is allowed to adjust the `rat` slot of any polynomial object to make it rational in this sense. If this were to be a common operation we would provide a function to do it.)

Rather than specifying polynomials directly by their coefficient vector it is sometimes useful to construct the monic polynomial of lowest degree with a specified set of zeros. This is allowed by the convenience function `polyWithZeros`.

```
> p <- polyWithZeros(-2:4)
> p
48*x  - 28*x^2 - 56*x^3 + 35*x^4 + 7*x^5 - 7*x^6 + x^7
> solve(p)
[1] -2+2.2152e-20i -1-4.4304e-20i  0+0.0000e+00i  1+0.0000e+00i
[5]  2+1.1076e-19i  3-1.3878e-19i  4+5.0176e-20i
```

The `solve` method (not shown here) for polynomial objects finds the zeros which will normally be complex, or at least have complex rounding error as here. Plotting the polynomial is another direct way of checking that it does have the specified zeros.

```
> plot(p, xlim = c(-2.1, 4.1))
> abline(h = 0, lty = 4)
> lines(deriv(p), lty = 2)
```

The result is also not shown, but we invite the reader to explore the facilities with this and other examples. The complete current code is in the scripts for this chapter.

# Chapter 6

# Using Compiled Code

A very important and powerful feature of the S environment is that it is not restricted to functions written in the S language but may load and use compiled routines written in C or FORTRAN, or perhaps other languages.[1] Moreover there are ways in which such externally written and compiled routines may communicate with the S session and make use of S functions and objects. In the new S engine and in R it is possible to manipulate S objects in C through the .Call interface.

## 6.1 Writing S functions to call compiled code

We have to specify both sides of the interface. For the S side it is assumed that the compiled code follows either C or FORTRAN calling conventions, although it need not have been written in either of those languages. The following protocol is used by the S functions called .C and .Fortran. (The function "%+%" on page 126 provides an example.)

- The first argument is a character string giving the name[2] of the routine.

- Each further argument must match the argument of the compiled routine. In particular the data passed through to the routine must have the correct storage.mode and must match the argument in length. Unlike S, neither FORTRAN nor C can deduce the length or mode of arguments.

- The arguments passed to the compiled routine may be given name fields. These do not match anything in the routine itself, but will be retained as name fields in the result.

- There can be further arguments that must come after all those that match the compiled routine (and must be named): these are NAOK, specialsok (not R), plus, for the old S engine, pointers, for the new S engine, COPY and CLASSES, and for R, DUP and PACKAGES.

---

[1] There are some versions of S-PLUS for which the facility is not supported or rather limited.
[2] this may need to be in all lower case for .Fortran, or in all upper case.

123

- The value returned by the call is a list containing all the arguments passed to the compiled code. The components of the list will reflect any changes made by the compiled code. Any attributes of the arguments will be retained so that arrays will be returned as arrays, for example.

The storage modes for arguments and their C and FORTRAN counterparts are given in Table 6.1. Note carefully that the integer and logical modes correspond to long in C in S-PLUS and int in R, and that these do differ on some 64–bit platforms.

Notice from Table 6.1 that the allowable argument types in C routines are all *pointers*. This is because the quantities manipulated are S vectors and so must be accessed by C indirectly. The case of character objects needs a little care. Recall that a character object is a *vector* of character strings; each string is an array of characters terminated by the ASCII character NUL (which has numeric code 0). This maps naturally to type char ** in C, but in FORTRAN character strings are stored as fixed-length one-dimensional character arrays, and the .Fortran interface allows at most a single character string to be passed, and whether even that works is compiler-dependent. Avoid FORTRAN character strings if at all possible.

S programmers can normally ignore the differences between storage modes double, single and integer, all of which correspond to S mode numeric. In the old S engine almost all[3] numeric objects were of storage mode double, but when using .C or .Fortran it is good practice to coerce the storage mode explicitly using the functions as.double, as.single and as.integer. Like as.vector these strip the attributes and return a vector; if preserving attributes is important it is necessary to use the assignment form of storage.mode:

```
storage.mode(x) <- "integer"
```

However, the attributes are copied to the result but they are not passed down to the compiled code: in particular matrices and arrays are passed as vectors in FORTRAN order, that is the first index varying fastest (see page 41 of MASS).

To illustrate the process consider a simple example. Figure 6.1 shows a C routine[4] to convolve finite sequences, that is to compute

$$c_i = \sum_{j,k,\geqslant 0: j+k=i} a_j b_k, \qquad i = 0, \ldots, n_a + n_b$$

A suitable driver routine for use with the old S engine and R would be

```
"%+%" <- function(a, b)
  .C("convolve",
     as.double(a), as.integer(length(a)),
     as.double(b), as.integer(length(b)),
     ab = double(length(a) + length(b) - 1))$ab
```

Since only the result component is needed, we only return that component. Note the use of the double function to generate a vector of storage mode double.

---

[3] the only common exceptions are sequences such as 1:10.

[4] Experienced C programmers will see ways to improve this code: but so will an optimizing compiler and in our experiments this form was as fast as any.

**Table 6.1**: Argument storage modes in S/R and corresponding data types for C and FORTRAN routines where applicable. (Based on Table 7.1 of Becker, Chambers & Wilks, 1988, p. 197 and Table 11.1 of Chambers, 1998, p. 412, with corrections.)

| S storage mode | C | FORTRAN |
|---|---|---|
| *For S-PLUS 3.x, 4.x and 2000:* | | |
| logical | long * | LOGICAL |
| integer | long * | INTEGER |
| double | double * | DOUBLE PRECISION |
| single | float * | REAL |
| complex | S_complex[a] * | DOUBLE COMPLEX |
| character | char ** | CHARACTER *(*)[b] |
| list | void ** | — |
| *For the new S engine:* | | |
| S *class* | | |
| logical | long * | INTEGER |
| integer | long * | INTEGER |
| numeric | double * | DOUBLE PRECISION |
| single | float * | REAL |
| complex | S_complex[a] * | DOUBLE COMPLEX |
| character | char ** | CHARACTER *(*)[b] |
| raw | char * | — |
| list | s_object ** | — |
| *For R:* | | |
| logical | int * | INTEGER |
| integer | int * | INTEGER |
| double | double * | DOUBLE PRECISION |
| or[c] | float * | REAL |
| complex | Rcomplex[d] * | DOUBLE COMPLEX |
| character | char ** | CHARACTER *255[b] |
| list | SEXP * | — |
| R object | SEXP | — |

---

[a] defined in S.h as a structure with two double components re and im. complex may also be used, but may cause conflicts and is proposed as a C reserved word.

[b] compiler-dependent, even on a single platform. It may be necessary to declare this as CHARACTER*n for a specific n.

[c] if marked by as.single, single or storage.mode <- "single".

[d] defined as a structure with two double components r and i.

```
void convolve(double *a, long *na,
              double *b, long *nb,
              double *ab)
{
  int i, j, nab = *na + *nb - 1;

  for(i = 0; i < nab; i++) ab[i] = 0.0;
  for(i = 0; i < *na; i++)
    for(j = 0; j < *nb; j++)
      ab[i + j] += a[i] * b[j];
}
```

**Figure 6.1**: File convolve.c: a C function to evaluate a discrete convolution operator. Replace long by int for R.

Supposing this function and the compiled code have been loaded into our copy of S-PLUS or R (see Appendix A) we can try them out. As we have given our function the name of a binary operator, we can use it by

```
> u <- rep(1, 5)
> u %+% u
[1] 1 2 3 4 5 4 3 2 1
> u %+% u %+% u
 [1]  1  3  6 10 15 18 19 18 15 10  6  3  1
```

Our libraries provide many useful examples of calls to C routines.

### Efficiency issues with the new S engine

The driver routine will also work with the new engine, but there are preferred styles that will make fewer copies of S objects. If the compiled code is used in only one place we can use

```
"%+%" <- function(a, b)
   .C("convolve",
      a, length(a), b, length(b),
      ab = double(length(a) + length(b) - 1),
      CLASSES = c("numeric", "integer", "numeric", "integer",
                  "numeric"),
      COPY = c(F, F, F, F, T))$ab
```

but if it is to be used in several places it is better to use

```
"%+%" <- function(a, b)
   .C("convolve",
      a, length(a), b, length(b),
      ab = double(length(a) + length(b) - 1))$ab
```

```
setInterface("convolve", "C",
  classes = c("numeric", "integer", "numeric", "integer",
              "numeric"),
  copy = c(F, F, F, F, T))
```

Both forms specify the classes of all of the arguments to the .C function, and the
actual arguments will be coerced to the required class if necessary. The COPY
or copy argument says that no copies are required as the compiled code does not
change the argument (such as a). The default behaviour is to make a copy and use
that copy as the argument, *if* the S object is used elsewhere. So in this example
the final argument will never be copied as the storage space allocated by the call
to double is not otherwise used. (The COPY or (especially) copy value should
reflect what the compiled code does, as the usage in the S code might be changed
at a later stage.)

The function unset provides another memory-saving device. The following
was from multinom in library nnet .

```
summ2 <- function(X, Y)
{
    X <- as.matrix(X); Y <- as.matrix(Y)
    n <- nrow(X); p <- ncol(X); q <- ncol(Y)
    Z <- t(cbind(X, Y))
    z <- .C("VR_summ2", n, p, q, Z = unset(Z), na = integer(1))
    Za <- t(z$Z[, 1:z$na, drop = F])
    list(X = Za[, 1:p, drop = F], Y = Za[, p + 1:q])
}
```

This says that although the C code overwrites the argument Z, no copy of the
original Z will be needed subsequently. (A simpler way to do this is to write
Z = t(cbind(X, Y)) in the call to .C.) Such efficiency tricks (including the
use of COPY ) should not be considered until speed is an issue.

## Issues with the R engine

In the S engines objects in the calling S functions are guaranteed not to move
during the current S function unless the object is assigned to. Thus we can pass
a pointer to an object down to C in one .C call and use the values pointed to in
subsequent .C calls. We use this in our nnet code, for example, to avoid making
a copy of the dataset.

R has garbage collection, which means that R objects can be moved in memory at any time the R engine is running (possibly during a .C call if call_R or
the allocation macros are used). This means that objects must be copied rather
than their pointers.

## 6.2   Writing compiled code to work with S

In this section we consider a number of issues in writing C or FORTRAN code to work with S-PLUS. Almost all the issues also apply to R.

It is possible to call routines within S-PLUS for memory allocation, to access the internal random number generator and to report warnings and errors. The routines we know of are given in Table 6.2. Declarations for these and other functions available to the C programmer are given in the include file S.h which resides in the include subdirectory of the S-PLUS or R home directory. Using Splus COMPILE or Splus SHLIB or R SHLIB will ensure that the correct include files are found.

We hide the differences in the calls between versions of S in a header file verS.h:

```
#if( defined(SPLUS_VERSION) && SPLUS_VERSION > 5100 )
#   define RANDIN   seed_in((long *)NULL, S_evaluator)
#   define RANDOUT seed_out((long *)NULL, S_evaluator)
#   define UNIF unif_rand(S_evaluator)
#elif( defined(SPLUS_VERSION) && SPLUS_VERSION >= 5000 )
#   define RANDIN   seedin((long *)NULL, S_evaluator)
#   define RANDOUT seedout((long *)NULL, S_evaluator)
#   define UNIF unif_rand(S_evaluator)
#else
#   define RANDIN   seed_in((long *)NULL)
#   define RANDOUT seed_out((long *)NULL)
#   define UNIF unif_rand()
#   define Salloc(n,t) (t*)S_alloc(n, sizeof(t))
#   define Srealloc(p,n,old,t) (t*)S_realloc(p,n,old,sizeof(t))
#   define S_EVALUATOR
#endif

#ifdef USING_R
  typedef double Sfloat;
  typedef int Sint;
# define SINT_MAX INT_MAX
#else
  typedef float Sfloat;
  typedef long Sint;
# define SINT_MAX LONG_MAX
#endif
```

Note the use of the macros USING_R which is defined in R but not in S, and SPLUS_VERSION which is only defined in S-PLUS.

For example, our spatial library uses

```
#include <S.h>
#include "verS.h"

void VR_pdata(Sint *npt, Sfloat *x, Sfloat *y)
```

**Table 6.2**: Symbols that can be used with the old S engine and R.

| | |
|---|---|
| `char *S_alloc(long n, int size)` | allocate n items of requested size and set them to zero. |
| `char *S_realloc(char *p, long new, long old, int size)` | reallocate new items of requested size, for pointer p to a block of size old. Allocated memory is zeroed. |
| `type *Calloc(int n, type)` | calloc n items of type type. |
| `type *Realloc(char * p, int n, type)` | realloc pointer p to n items of type type |
| `Free(char *p)` | 'free' memory pointed to by p. |
| `double unif_rand()` | one uniform random number. |
| `double norm_rand()` | one standard normal variate. |
| `setseed(long *seed)` | as in set.seed (not R). |
| `seed_in(long *iseed)` | set the seed. iseed should be either a pointer to a vector of 12 integers between 0 and 63, or NULL, when the seed is read in from .Random.seed. R ignores iseed and always reads in from .Random.seed. |
| `seed_out((long*)NULL)` | write the seed back out. |
| `PROBLEM ... RECOVER` | Equivalent of stop. See page 134. |
| `PROBLEM ... WARNING` | Equivalent of warning. |

There are some changes in the new S engine. The forms to be used there are

```
type *Salloc(long n, type)
type *Srealloc(char *p, long new, long old, type)
double unif_rand(S_evaluator)
double norm_rand(S_evaluator)
seedin(long *iseed, S_evaluator)
seedout((long*)NULL, S_evaluator)
```

The last two apply to S-PLUS 5.0 and 5.1. Future versions will use

```
seed_in(long *iseed, S_evaluator)
seed_out((long*)NULL, S_evaluator)
```

```
{
    S_EVALUATOR
    int      i;
    Sfloat   ax, ay;

    testinit();
    ax = xu0 - xl0; ay = yu0 - yl0;
    RANDIN;
    for (i = 0; i < *npt; i++) {
        x[i] = xl0 + ax * UNIF; y[i] = yl0 + ay * UNIF;
    }
    RANDOUT;
}
```

which will work on any of the systems. The macro S_EVALUATOR is needed only if the symbols in Table 6.2 are used.

It is probably best to avoid these calls if possible: for example almost always the random-number seed can be set in the calling S code. In this example the random-number generation could have been done in S and passed to C, but in many similar examples the number of random numbers required is not known in advance. If random number generation is to be done in C, it is essential to call the operations represented by the macros RANDIN and RANDOUT to read in the initial seed from .Random.seed and write out the final seed.

## Allocating storage

If the C routine dynamically allocates storage it should use S_alloc which is similar to malloc except that the space is automatically freed afterwards (even if S errors occur). Note that the call is

```
S_alloc(long n, int size)
```

like calloc rather than malloc and, like calloc, the allocated memory is zeroed. There is also S_realloc which is similar to realloc and again zeroes the allocated storage. (There are differences in the new S engine: see Table 6.2.) These routines raise S errors with messages if allocation fails.

The documentation on *when* S-PLUS frees the space is confused. The Programmer's Guides say:

> The storage they allocate lasts until the current evaluation frame goes away
> (at the end of the function calling .C() ) ...

but that is not usually correct. The storage will always last until the end of the call to .C; some .C calls do not generate an evaluation frame and so the storage will last until the end of the calling function, but for most calls storage will go away at the end of the .C call. (The rules governing which calls do not generate an evaluation frame are complex and this should not be relied upon.)

In R the memory is always released at the end of the call to .C.

There are also functions `Calloc`, `Realloc` and `Free` to replace the usual un-capitalized versions, which have the two advantages of returning S-style error messages and of casting the pointers to the desired type. Memory allocated by these calls lasts until it is explicitly freed by `Free`. A call to S_realloc can re-allocate a null pointer, but `Realloc` can not.

Since this area is system-dependent users may need to consult their implementation literature for further details.[5]

*Examples*

These memory allocation routines are widely used in our libraries. In library `spatial` we have

```
static Sfloat *alph1=NULL;

void VR_alset(Sfloat *alph, Sint *nalph)
{
    int i;

    if(alph1 != NULL) alph1 = Realloc(alph1, *nalph, Sfloat);
    else alph1 = Calloc(*nalph, Sfloat);
    for (i = 0; i < *nalph; ++i) alph1[i] = alph[i];
}
```

This memory is never freed but re-used on each call to a kriging routine. It replaces an earlier static allocation

```
static Sfloat alph1[10002];
```

which was wasteful (the length of `alph` was normally about 1000) and provided a hard limit that the current code avoids.

Many of our functions make one call to `.C` to pass in data, then more to manipulate it. Then the `Calloc` family must be used, with a final call to `Free` to free up the storage. For example, `isoMDS` in library `MASS` uses

```
void
VR_mds_init_data(Sint *pn, Sint *pc, Sint *pr, Sint *orde,
                 Sint *ordee, double *xx)
{
    int i;

    n = *pn; nr = *pr; nc = *pc; dimx = nr * nc;
    ord = Calloc(n, Sint); ord2 = Calloc(n, Sint);
    x = Calloc(dimx, double); d = Calloc(n, double);
    y = Calloc(n, double); yf = Calloc(n, double);
    for (i = 0; i < n; i++) ord[i] = orde[i];
    for (i = 0; i < n; i++) ord2[i] = ordee[i];
    for (i = 0; i < dimx; i++) x[i] = xx[i];
}
```

---

[5]We are grateful to Bill Dunlap of MathSoft DAPD for clarifying many of the details for us.

```
void
VR_mds_unload()
{
  Free(ord); Free(ord2); Free(x); Free(d); Free(y); Free(yf);
}

void
VR_mds_dovm(double *val, Sint *maxit, Sint *trace, double *xx)
{
  int     i;
  vmmin(dimx, x, val, (int) *maxit, (int) *trace);
  for (i = 0; i < dimx; i++) xx[i] = x[i];
}
```

called as (in the old S-PLUS engine or in R)

```
    ....
on.exit(.C("VR_mds_unload"))
.C("VR_mds_init_data",
    as.integer(nd), as.integer(k), as.integer(n),
    as.integer(ord - 1), as.integer(order(ord) - 1),
    as.double(y))
tmp <- .C("VR_mds_dovm",
            val = double(1), as.integer(maxit),
            as.integer(trace), y = as.double(y))
    ....
```

If all the calculations are done in one call to .C, as in our function
lqs.default, we can use

```
#include <S.h>
#include "verS.h"

static void lqs_setup(Sint *n, Sint *p, Sint *ps)
{
  coef = Salloc(*p, double);
  qraux = Salloc(*p, double);
  work = Salloc(2*(*p), double);
    ....
}
    ....
```

and the storage will automatically be released when the call to lqs finishes.

## Using C input/output

There is a problem with the use of the usual I/O using stdin, stdout and
stderr on S-PLUS 4.x under Windows. To work around this portably, include
the following in your C code

```
#include <S.h>
#if defined(SPLUS_VERSION) && SPLUS_VERSION >= 4000
#   include <newredef.h>
#endif
```

Without this, the I/O is delayed or lost. Further, `scanf` cannot be used; use `fgets` and `sscanf` instead.

Although this is not strictly needed for S-PLUS 5.x, it is recommended.

Standard I/O cannot be used with the console version of R under Windows. Replace calls to `printf` by `Rprintf`, and to `fprintf(stderr, ...)` by `REprintf(...)`.

Beware of assuming that there is a user sitting at a console with whom to interact: with the increasing use of inter-process communications it is quite possible that input and output involve a link to another process. If possible, use S code for all the input and output.

FORTRAN input/output

S-PLUS does not use FORTRAN input/output routines, and these may not work correctly if included in your FORTRAN code. Instead, the output functions DBLEPR, INTPR and REALPR are provided.[6] These all have the form

```
SUBROUTINE name(LABEL,NCHAR,DATA,NDATA)
```

where LABEL is a character string used for the printout, NCHAR is the length of the label (or zero for no label) and there are NDATA items to be printed from array DATA. The names correspond to the FORTRAN data types of DOUBLE PRECISION, INTEGER and REAL. These functions often suffice, especially for debugging. If elaborate layouts are needed, information can be written to a character string, and any of these routines used to print the string as a label. (Thus there is no need for 'CHARPR'.) When using R and on *some* S platforms NCHAR can be set to −1 when the length of the label will be computed internally. This is not portable whereas the FORTRAN function LEN can always be used to compute the length.

FORTRAN input/output can be used with R, but the xxxxPR subroutines are available and are convenient for debugging. Note that, depending on the compilers, there can be problem with buffering if both C and FORTRAN I/O are used, and the output can appear in a fairly arbitrary order unless the buffers are flushed (which may not be possible from FORTRAN). This is avoided by using the xxxxPR functions which employ the C output routines, and as such are the only means of I/O guaranteed to work on GUI versions of R.

---

[6]These do rely on using compatible C and FORTRAN compilers.

## Using C++

On most platforms it is relatively straightforward to use C++ code. There are two issues, how to interface the code (discussed here) and how to ensure that the libraries and initialization code get invoked, discussed in Appendix A.

It will normally be easiest to write a wrapper for C++ code to be linked to S, for example to convert from the pointer types used by .C to the C++ classes. The parts of that wrapper that define symbols to be called from .C need to be wrapped in an extern "C" declaration. Normally that is all that is required.

More effort may be needed to use C++ input/output, as the C++ iostream library will normally use a different set of buffers from C (and from FORTRAN) I/O, and C++ I/O will appear out of order with the C-based I/O from the S engine. One possible work-around is to call ios::sync_with_stdio() if your version of iostream supports it. However, this will not help if standard C I/O does not work (as in the GUI-based systems), so if possible avoid C++ I/O.

## Handling errors

A number of authors have left calls to exit (C) or STOP (FORTRAN) in their compiled code. This is anti-social, as such calls will cause S-PLUS to terminate, not just the author's code. Fortunately, more elegant ways have been provided (and documented).

### *From* C

The mechanisms

```
#include <S.h>
PROBLEM "warning message" WARNING(NULL_ENTRY);
PROBLEM "error message" RECOVER(NULL_ENTRY);
```

can be used to emulate calls to warning and stop respectively. The unusual syntax covers macro calls to sprintf, so the string can be replaced by any set of arguments to that function, separated by commas. Thus we could use

```
PROBLEM "#params (%d) is too large", nparams
   RECOVER(NULL_ENTRY);
```

To use these with S-PLUS 4.0 release 3, 4.5 and 2000, add the compiler flag

```
-DS_ENGINE_BUILD
```

when compiling for use with dynamic or static loading (but not DLL loading).

Once again, it is often helpful to hide the details. For example, the function errmsg used on page 136 is defined as

```
static void errmsg(char *string)
{ PROBLEM "%s", string RECOVER(NULL_ENTRY); }
```

With R this can be replaced by a direct call to the C function error.

In the new S engine the macro RECOVER(NULL_ENTRY) can be replaced by ERROR, WARNING(NULL_ENTRY) by WARN, and there is also a macro PRINT_IT.

On some engines these macros use a fixed size buffer (around 4Kbytes), so error messages should not be too verbose.

*From* FORTRAN

S-PLUS (but not R) defines a function XERROR called as

```
CALL XERROR('msg', LEN('msg'), errno, errlev)
```

where errlev is the error level ( 2 for a fatal error, 0 for a warning, −1 for a warning to be given just once) and errno is a strictly positive identifier for that error message, unique within that subroutine. See the help page for further details and the similar function XERRWV .

## Missing and special values

Internally S-PLUS and R work with missing values ( NA) and (on most platforms) the special values of IEEE floating point arithmetic ( Inf , −Inf and NaN, the latter denoting an indefinite result like $0./0.$). By default an error is generated if either occur in an argument passed to .C or .Fortran .

When C or FORTRAN code is being used for speed, it may be important to handle the missing and special values in the compiled code. The arguments NAOK and specialsok (not R) to .C or .Fortran can be set to true, in which case the S calling routine performs no checking, so the user's code must check *all* the arguments. We describe how to do this below; the details vary between S-PLUS and R.

No special facilities are provided for handling missing values in FORTRAN code. Often the simplest way to handle missing values (even with C) is to pass to the compiled code a vector giving is.na(x) . This idea of making the S computation do the work is a very good one unless very large objects are to be handled.

S-PLUS *macros*

S-PLUS provides a set of C macros[7] to test and set missing and special values. These include

```
is_na(x, mode)   is_nan(x, mode)        na_set(x, mode)
is_inf(x, mode)  inf_set(x, mode, sign) na_set3(x, mode, type)
```

Here x is a *pointer* to a numeric type, and mode is one of the symbolic constants INT, REAL, DOUBLE, COMPLEX, CHAR or LGL (for logical). Note that NaN is treated as a type of missing value; the function is_na returns 0, Is_NA or Is_NaN and the type argument to na_set3 should be one of these two symbolic constants. IEEE infinite values are signed, so is_inf returns 0 or ±1.

As an example, consider rewriting the function dist . This operates on an $n \times p$ matrix, and for each pair of rows $i, j$ computes the distance omitting columns in which either row has a missing value, and rescaling appropriately. The S function is

---

[7]These should be used in the new S engine, not those defined on page 425 of Chambers (1998).

```
ourdist <- function(x)
{
    n <- nrow(x)
    res <- .C("ourdist",  as.double(x), as.integer(n),
              as.integer(ncol(x)),  res = double(n*(n-1)/2),
              NAOK = T)$res
    attr(res, "Size") <- n
    res
}
```

and the (unoptimized) C code is

```
/* S-PLUS version */
#include <S.h>
#include <math.h>

void  ourdist(double *x, long *nin, long *pin, double *res)
{
    int     i, j, k, den, n = *nin, p = *pin;
    double  item, tmp;

    if(is_na(&n, INTEGER) || n < 2)
        errmsg("n is zero, one  or missing")
    for (i = 0; i < n-1; i++)
        for (j = i+1; j < n; j++) {
            den = 0; tmp = 0.0;
            for (k = 0; k < p; k++)
                if (!is_na(x + i + n * k, DOUBLE) &&
                    !is_na(x + j + n * k, DOUBLE)) {
                    den++;
                    item = x[i + n * k] - x[j + n * k];
                    tmp += item * item;
                }
            if (!den) na_set3(&tmp, DOUBLE, Is_NaN);
            else tmp = sqrt(tmp * p / den);
            *res++ = tmp;
        }
}
```

This works well provided the missing values really are NA s and not NaN s
(and they will print exactly the same, see page 22). Otherwise you will see an
error message like

```
Problem in .C("ourdist",: subroutine ourdist: 1 NaN value(s) in
argument 1
```

If we change the call to include specialsok=T our routine will then work correctly since !is_na covers NA and NaN . However, this also allows infinite values
to be passed in. A little thought shows that in our example this can only result
in distances that are +Inf or NaN and the calculations should be correct, at least
on machines which perform IEEE floating-point calculations. Otherwise is_inf
and inf_set will need to be used to set tmp appropriately.

**R** *macros*

R provides a different set of C macros. These include ISNA, ISNAN and FINITE
to be applied to doubles (not pointers), and the values NA_REAL, NA_INTEGER,
NA_LOGICAL, NA_STRING, R_PosInf and R_NegInf that can be used in assign-
ments. In R NaN s really do print as NaN, but NAOK = TRUE allows NaN s and
Inf s through as well as NA s. In converting ourdist.c we have to note that
ISNA does not detect NaN s but ISNAN is true for NA as well as NaN, so we need

```
/* R version */
#include "R.h"   /* Please check this for your R version */
#include <math.h>

void  ourdist(double *x, int *nin, int *pin, double *res)
{
    int      i, j, k, den, n = *nin, p = *pin;
    double   item, tmp;

    if(n == NA_INTEGER || n < 2)
        errmsg("n is zero, one  or missing")
    for (i = 0; i < n-1; i++)
        for (j = i+1; j < n; j++) {
            den = 0; tmp = 0.0;
            for (k = 0; k < p; k++)
                if (!ISNAN(x[i + n * k]) && !ISNAN(x[j + n * k]))
                {
                    den++;
                    item = x[i + n * k] - x[j + n * k];
                    tmp += item * item;
                }
            if (!den) tmp = R_NaN;
            else tmp = sqrt(tmp * p / den);
            *res++ = tmp;
        }
}
```

The FINITE macro is used quite often, as it is false for all the special values (NA,
NaN, Inf, -Inf).

## Calling FORTRAN from C

It is quite common to want to call FORTRAN routines from C routines to be compiled into S-PLUS or R; perhaps the commonest use is to call numerical analysis
routines either in separately compiled FORTRAN code or those already included.
Most package writers just call the FORTRAN name with a trailing underscore; this
works on most platforms, but not all. The portable approach is to use a set of
macros defined by including S.h in the C code.

Here is an example from our lqs.default function for S-PLUS:

```
/* find qr decomposition, basis of qr.default() */
void F77_NAME(dqr)(double *qr, long *dx, long *pivot,
                   double *qraux, double *tol, double *work,
                   long *rank);

/* solve for coefficients */
void F77_NAME(dqrsl)(double *qr, long *ldx, long *n, long *rank,
                     double *qraux,
                     double *y1, double *d1, double *y2,
                     double *coef, double *d3, double *d4,
                     long *job, long *info);
  ....
  /* compute fit, find residuals */
  dx[0] = nnew; dx[1] = *p;
  for(i = 0; i < *p; i++) pivot[i] = i+1;
  rank = *p;
  F77_CALL(dqr)(xr, dx, pivot, qraux, &tol, work, &rank);
  if(rank < *p) { (*sing)++; continue; }
  F77_CALL(dqrsl)(xr, &nnew, &nnew, &rank, qraux, yr, &dummy,
                  yr, coef, &dummy, &dummy, &n100, &info);
```

The prototypes for the forward declaration were found by examining the FOR-
TRAN code and the types of other calls. Note that different macros are needed
for declarations and for the actual calls, and that the FORTRAN functions *must* be
declared.

There are similar macros F77_COMDECL and F77_COM to declare and use a
FORTRAN common block. To write a C function which is callable from FOR-
TRAN use the macro F77_SUB for the declaration of the name: this will append
an underscore on systems where this is needed. On most current platforms all the
F77_* macros are the same, but not quite on all, so they should be differentiated
carefully. To summarize

Use F77_NAME to declare a FORTRAN subroutine.

Use F77_CALL to call it.

Use F77_COMDECL to declare a FORTRAN common block in C.

Use F77_COM to use variables in the common block.

Use F77_SUB to define a C function to be called by FORTRAN.

Mixing C and FORTRAN only works well if compatible compilers are used.
Using the f2c FORTRAN to C translator[8] on the FORTRAN component of a
mixed-language project is often a more portable route.

## 6.3  Calling S from C

As well as calling C routines from S it is also possible to do the inverse operation
of calling an S function from a C routine with the C routine call_S provided

---

[8] available from netlib.att.com and mirrors.

by S. This is necessary, for example, in a routine for numerical integration where the integrand may be an S function. Note that it is only possible to use `call_S` to call S functions from within C functions called from S. The process may be extremely slow. (It is often more convenient to generate a look-up table from the S function and pass that to the C or FORTRAN function. This approach was tried for `surf.gls` in our library `spatial` and was several times faster and so was adopted.)

The `.Call` interface (Section 6.4) provides an alternative mechanism for manipulating and evaluating S expressions. This is the preferred approach where it is available.

### S-PLUS version

The arguments for `call_S` are

```
void call_S(void *func, long nargs, void **arguments,
            char **modes, long *lengths, char **names,
            long nres, void **results);
```

Here `func` is the S code that is passed to C as a pointer to a list, the next five arguments give the number of arguments of the S function, and pointers to their (vector) values, modes, lengths and (optionally) names. The variable `nres` describes the number of components of the result, and `results` is an array of pointers that the `call_S` interface sets to storage where the results have been allocated.

There are very few examples available[9] of the use of `call_S`, so a detailed example may be helpful. The essential part of the C code for computing covariances in our `spatial` library was

```
#include <S.h>
#include <math.h>
#define max(a,b) ((a > b)? a:b)

static double eps;
static void *Scovmod;

void VR_cmset(double *ineps, void **Scode)
{
  eps = *ineps; Scovmod = *Scode;
}

static void cov(long n, double *d, long pred)
{
  int i;
  char *modes[] = {"double"};
  long lengths[1];
  void *arguments[1], *results[1];
```

---

[9]there is one in Becker *et al.* (1988) and another in the S-PLUS Programmer's Guide.

```
    for (i = 0; i < n; ++i) d[i] = max(sqrt(d[i]), pred*eps);
    arguments[0] = (void *)d; lengths[0] = n;
    call_S(Scovmod, 1L, arguments, modes, lengths,
           (char **)NULL, 1L, results);
    for (i = 0; i < n; ++i) d[i] = ((double *)(*results))[i];
}


void test_cov(long *n, double *d, long *pred)
{
   cov(*n, d, *pred);
}
```

and a test S function might be

```
testcov <- function(x, covmod, ...)
{
    n <- nrow(x)
    covmod <- covmod
    args <- list(...)
    if(length(args)) {
       pm <- pmatch(names(args), names(covmod), nomatch=0)
       if(any(pm == 0))
           warning(paste("some of ... do not match"))
       covmod[pm] <- unlist(args)
    }
    defn <- c("function(x){covmod <-", deparse(covmod),
              "covmod(x)}")
    krcov <- eval(parse(text=defn))
    d <- c(0, dist(x))
    eps <- 1e-3 * max(d)
    .C("VR_cmset", as.double(eps), list(krcov))
    z <- .C("test_cov", as.integer(1 + n*(n-1)/2),
            cov = as.double(d), as.integer(1))$cov
    cov <- matrix(0, n, n)
    cov[lower.tri(cov)] <- z[-1]
    cov <- cov + t(cov)
    diag(cov) <- z[1]
    cov
}
```

This constructs a function krcov which takes a single argument and returns a vector of the same length. The specified covariance function is included as a local function within krcov to ensure that it is fully specified when stored. The second argument of the call to VR_cmset then passes a pointer to this function to the C code which is used to stored the code's address in Scovmod.

We can test this by

```
> library(spatial)
> x <- matrix(rnorm(12),,2)
> testcov(x, expcov, d=0.7, se=5)
```

```
          [,1]    [,2]    [,3]    [,4]    [,5]    [,6]
[1,] 24.9983  2.8148  2.1166  3.6172  1.5298  1.1402
[2,]  2.8148 24.9983  3.1736  4.8294  2.9826  2.5907
[3,]  2.1166  3.1736 24.9983  5.1552  5.5708  2.6656
[4,]  3.6172  4.8294  5.1552 24.9983  3.3208  2.1194
[5,]  1.5298  2.9826  5.5708  3.3208 24.9983  4.5665
[6,]  1.1402  2.5907  2.6656  2.1194  4.5665 24.9983
```

### R version

R has an interface call_R that is almost but not quite identical. The types of the arguments are:

```
void call_R(char *func, long nargs, void **arguments,
            char **modes, long *lengths, char **names,
            long nres, char **results)
```

but the main difference is how the S function is passed down by .C, as void * and not wrapped in a list. As R now has a .Call interface, we recommend not using call_R in new projects. (A call_R version of our example is given in the on-line scripts for this chapter.)

## 6.4  Using the .Call interface

In the new S engine it is possible to manipulate S objects in C code through the .Call interface. Later in this section we consider the .Call and .External interfaces in R, which provide equivalent functionality.

Chambers (1998, Appendix A) gives somewhat fuller details. It is clear that using .Call is complex and has considerable traps for programmers who are less than fastidious, and there are no tools (like lint) to check the details automatically.

Let us illustrate the use of .Call by re-writing our convolution operator %+% once again. The S part is easy: we just pass in the S objects as arguments to .Call and return the result, as in

```
"%+%" <- function(a, b)
  .Call("convolve2", as.double(a), as.double(b))
```

We simplify the C code by coercing the arguments in the S code.

We then have to write the C function, compile it and load it into S-PLUS. We used

```
#include <S.h>

s_object * convolve2(s_object *a, s_object *b)
{
  int i, j;
  long na, nb, nab;
```

```
double *xa, *xb, *xab;
s_object *ab;

S_EVALUATOR
na = LENGTH(a); nb = LENGTH(b); nab = na + nb - 1;
ab = NEW_NUMERIC(nab);
xa = NUMERIC_POINTER(a); xb = NUMERIC_POINTER(b);
xab = NUMERIC_POINTER(ab);
for(i = 0; i < nab; i++) xab[i] = 0.0;
for(i = 0; i < na; i++)
  for(j = 0; j < nb; j++)
    xab[i + j] += xa[i] * xb[j];
return(ab);
}
```

To understand this, note that S objects are (understandably) of type s_object. The first line, S_EVALUATOR, is always needed: it is a macro and not terminated by a semicolon. The next few lines are the equivalent of

```
na <- length(x), nb <- length(b);  nab <- na + nb - 1;
ab <- numeric(nab)
```

Then the NUMERIC_POINTER lines extract pointers to the data from the S objects for manipulation as C arrays. Finally we manipulate the data pointed to and return (a pointer to) the object we created.

So far so good. There are macros NEW_*type*, AS_*type*, IS_*type*, *type*_POINTER and *type*_VALUE (which returns a single value) to each of the S atomic types (NUMERIC, SINGLE, INTEGER, CHARACTER, LOGICAL and RAW as well as LIST) so we can do manipulations of this sort. The difficulty is that we as C programmers have taken over responsibility for error checking. For example, the code shown has no check that a is numeric, and if we had not done the coercion in the S code we would have needed to add lines like

```
if(!(a = AS_NUMERIC(a))) errmsg("a is not numeric");
if(!(b = AS_NUMERIC(b))) errmsg("b is not numeric");
```

Note too that it is important to do the coercion, as the data part of an object inheriting from numeric could be very different. NUMERIC_POINTER does not check if it is given a numeric object: the checking is the programmer's responsibility.

That responsibility becomes heavier if we alter an existing S object (which does not happen in our example) because of shared references to objects. It is the job of the C program to make a copy as necessary before altering the (copied) object. For atomic arguments this is done by the .Call wrapper *unless* the COPY argument is used (see page 126) or setInterface is invoked. However, for recursive objects like lists this only copies the top set of pointers, not what they point to. The C program can use the macros COPY or COPY_ALL to copy the current level (all of an atomic object) or the whole object, and the macro

```
SET_ELEMENT(x, i, value)
```

should be used for x[[i + 1]] <- value (note the use of C–style 0–based indexing *versus* S–style 1–based indexing).

## Evaluating S expressions

It is possible to evaluate arbitrary S expressions using the macros `EVAL` and
`EVAL_IN_FRAME`. An expression is just an S object `expr`, and can be evaluated either in the local frame by `EVAL(expr)` or in frame n (a number) by
`EVAL_IN_FRAME(expr, n)`. (The frame number will typically have been passed
in after calculations involving `sys.parent` and related functions.)

Objects can be retrieved or assigned by

```
char *name; long n; s_object *value;

GET(name)                    GET_FROM_FRAME(name, n)
ASSIGN(name, value)          ASSIGN_IN_FRAME(name, value, n)
```

where the first versions use the local frame. As the objects here are handled by S,
it will make the necessary copies on assignment.

We often want to examine parts of an S object, so there are also macros such
as

```
GET_DIM(x)                   SET_DIM(x, d)
GET_DIMNAMES(x)              SET_DIMNAMES(x, dimnames)
GET_ROWNAMES(x)              GET_COLNAMES(x)
GET_NAMES(x)                 SET_NAMES(x, names)
GET_LEVELS(x)                SET_LEVELS(x, levels)
GET_LENGTH(x)                GET_ATTR(x, "what")
GET_CLASS(x)                 SET_CLASS(x, class)
GET_CLASS_NAME(x)
```

the last giving the name as a character vector rather than an S object.

*An example*

Let us consider how the spatial covariance calculation for which we used `call_S`
could be done with `.Call`.

```
testcov <- function(x, covmod, ...)
{
    ....
    z <- .Call("Call_cov", as.double(d), as.double(eps),
               functionBody(krcov), new.frame())
    ....
```

where the C code is

```
#include <S.h>
#include <math.h>
#ifndef max
#  define max(a,b) ((a > b)? a:b)
#endif

s_object * Call_cov(s_object *d, s_object *eps,
```

```
                      s_object *covexpr, s_object *frame)
{
  s_object *cov;
  int i, nframe = INTEGER_VALUE(frame);
  double *xd, xeps;

  S_EVALUATOR
  xd = NUMERIC_POINTER(d); xeps = NUMERIC_POINTER(eps)[0];
  for (i = 0; i < LENGTH(d); ++i)
    xd[i] = max(sqrt(xd[i]), xeps);
  ASSIGN_IN_FRAME("x", d, nframe);
  cov = EVAL_IN_FRAME(covexpr, nframe);
  return(cov);
}
```

Note that this is not the equivalent of a call to `call_S`: rather we can use a `.Call` call to replace a `.C` call, and use `EVAL` or `EVAL_IN_FRAME` to replace the call to `call_S` in the C code. In a more complete replacement for `surf.gls` d would be constructed in the C code and `cov` used in the C code by something like

```
double *xcov;
s_object *d;

/* given C array xd[n] */
d = MAP_NUMERIC(xd, n, NULL);
ASSIGN_IN_FRAME("x", d, nframe);
cov = EVAL_IN_FRAME(covexpr, nframe);
xcov = NUMERIC_POINTER(cov);
/* use xcov[n] as a C array */
```

### Interfaces in R

R has `.External` and `.Call` interfaces that allow S objects to be passed to C code for manipulation. They differ in the way the arguments are passed: `.Call` passes them separately, and `.External` passes them as an single SEXP (a pointer to an R internal structure). As with the new S interface, the C code manipulates through an extensive set of macros, which are defined in the header file `Rinternals.h`. There are extensive examples of the use of these macros in the public R source code and in some user-contributed packages. There is also a compatibility header `Rdefines.h` that allows many of the macros from the new S engine to be used.

There is fairly extensive documentation in the R manuals.

Let us show how we coded our convolution example, first using the `.Call` interface, then the `.External` interface.

```
#include <Rinternals.h>

SEXP convolve3(SEXP a, SEXP b)
{
```

```
    int i, j, na, nb, nab;
    SEXP ab;

    na = length(a); nb = length(b); nab = na + nb - 1;
    PROTECT(ab = allocVector(REALSXP, nab));
            /* or ab = NEW_NUMERIC(nab) */
    for(i = 0; i < nab; i++) REAL(ab)[i] = 0.0;
    for(i = 0; i < na; i++)
      for(j = 0; j < nb; j++)
        REAL(ab)[i + j] += REAL(a)[i] * REAL(b)[j];
    UNPROTECT(1);
    return(ab);
}
```

We can also use the `NUMERIC_POINTER` approach shown in `convolve2`. Note that we have omitted error-checking: all the necessary checking is done by the R functions.

Note the use of the `PROTECT` and `UNPROTECT` macros. These handle the protection of objects from R's garbage collection. Here we ask that (the memory pointed to by) `ab` not be released until we have finished with it; a call to `UNPROTECT(n)` removes the protection from the last `n` objects that were protected. Generally, we need to use `PROTECT` to tell R that we are using an object so that R knows it is in use and will retain it and update the pointer to it if it is moved. Note that we need to say we have finished with the object before returning it, but that is safe as immediately afterwards R will mark it as used by R. Arguments are already in use and hence protected.

The user is responsible for housekeeping the `PROTECT` calls, which should all be cleared before exiting the function, including on error conditions. (Calling `PROBLEM ... RECOVER` will clear all the current PROTECTions.) R will give a run-time warning if the housekeeping is incorrect. A more serious error is the omission of `PROTECT` calls. Note that it is almost impossible to check this empirically, as garbage collection is essentially random. Use of `gctorture` will maximize the use of garbage collection and so help to 'ferret out memory protection bugs'. At present protection of `ab` is unneeded in this example, but unless the programmer knows precisely which macros might trigger garbage collection it is safest to protect.

We can call this code when compiled by

```
"%+%" <- function(a, b)
  .Call("convolve3", as.double(a), as.double(b))
```

To use the `.External` interface the R side is almost unchanged

```
"%+%" <- function(a, b)
  .External("convolve4", as.double(a), as.double(b))
```

but the C code has a single argument of type `SEXP` that needs to be unpicked. We illustrate one style of doing so in `convolve4`.

```
#include <Rinternals.h>

SEXP convolve4(SEXP args)
{
  int i, j, na, nb, nab;
  SEXP a, b, ab;

  args = CDR(args); a = CAR(args);
  args = CDR(args); b = CAR(args);
  na = length(a); nb = length(b); nab = na + nb - 1;
  PROTECT(ab = allocVector(REALSXP, nab));
  for(i = 0; i < nab; i++) REAL(ab)[i] = 0.0;
  for(i = 0; i < na; i++)
    for(j = 0; j < nb; j++)
      REAL(ab)[i + j] += REAL(a)[i] * REAL(b)[j];
  UNPROTECT(1);
  return(ab);
}
```

This uses the CDR macro[10] to advance along the list of arguments, and the CAR macro to extract the argument. The shortcuts

```
first = CADR(args);
second = CADDR(args);
third = CADDDR(args);
fourth = CAD4R(args);
```

provide a convenient way to pick off up to four arguments. This we could have written

```
a = CADR(args); b = CADDR(args);
```

The two styles can be mixed as the shortcuts do not change args.

*Evaluating S expressions*

Evaluation of S expressions is done by the C function eval, whose arguments are an expression and an environment frame in which to evaluate the expression. Our testcov example can be re-written for R by

```
testcov <- function(x, covmod, ...)
{
    ....
    z <- .Call("RCall_cov", d, eps, body(krcov), new.env())
    ....
```

where the C code is

---

[10]The terminology is borrowed from LISP.

```
#include <Rinternals.h>
#include <math.h>
#ifndef max
#  define max(a,b) ((a > b)? a:b)
#endif

SEXP RCall_cov(SEXP d, SEXP eps, SEXP covexpr, SEXP rho)
{
  SEXP cov;
  int i;
  double *xd = REAL(d), xeps = REAL(eps)[0];

  for (i = 0; i < length(d); ++i)
    xd[i] = max(sqrt(xd[i]), xeps);
  defineVar(install("x"), d, rho);
  cov = eval(covexpr, rho);
  return(cov);
}
```

There are a couple of new features here. Function `defineVar` sets the value of the specified variable in the given environment. This is the equivalent of `assign` with `inherits = FALSE`; `setVar` is the equivalent with `inherits = TRUE`. Second, to refer to an R object by name, we need to look up its name in the symbol table. Function `install` does that lookup as a by-product of its main purpose, which is to put the name in the symbol table if it is not already there.

To get objects, use `findVar` or, for functions, `findFun`.

*Handling* ... *arguments*

It is not usually a good idea to handle a variable number of arguments in the C code, and it is never necessary, as one can just wrap ... by `list`, as in

```
limits <- function(..., na.rm=TRUE)
  .Call("limits", list(...), na.rm)
```

Nevertheless, it is possible to process a variable number of arguments in R using the `.External` interface. As an illustration, look at

```
> showargs <- function(...) invisible(.External("showArgs", ...))
> showargs(a=1, 2:5)
[1] 'a' length 2 1.000000 ...
[2] '' length 4 2 ...
```

where the C code is

```
#include <Rinternals.h>

SEXP showArgs(SEXP args)
{
  int i = 1, len;
  Rcomplex cpl;
```

```
    char *name;

    args = CDR(args);
    while(args != R_NilValue) {
      name = CHAR(PRINTNAME(TAG(args)));
      len = length(CAR(args));
      switch(TYPEOF(CAR(args))) {
      case REALSXP:
          printf(" [%d] '%s' length %d %f ...\n", i+1, name,
                    len, REAL(CAR(args))[0]);
          break;
      ....
      }
      args = CDR(args);
    }
    return(R_NilValue);
}
```

This scans through an arbitrary number of arguments, and prints a few details on each, purely for illustration and omitting error-checking.

## 6.5   Debugging compiled code

Simple debugging can often be performed by adding output statements to the code, in C or in FORTRAN. Remember that with FORTRAN under S-PLUS DBLEPR and related functions should be used, and for C code used with S-PLUS 4.x newredef.h *must* be included. Indeed, in some environments this is the only way to debug compiled code. It is often helpful to flush the output buffers after each message is written, so messages are seen before a program crash.

A symbolic source-code debugger can be used under certain circumstances. S-PLUS is not compiled for debugging, so the only circumstances known to us under which compiled code can be debugged is when using static loading or a DLL under Windows. The latter is particularly convenient when using Visual C++'s development environment: just build a debug version of the DLL and set the full path to the S-PLUS executable as the executable to be run under debug. Then set breakpoints in the source code for the DLL and go.

As R's source code is available, a debug version can be compiled by setting appropriate compiler flags at build time. However, there is still the problem of how to set breakpoints in your own code, as that will not be loaded when R is run under the debugger. One way out is to run, enter the commands to load the shared library at the S prompt, then interrupt and set breakpoints as required. The interruption is not so easy under Windows, but instructions are given in the R for Windows FAQ.

It can be much easier to debug the code while called from a C or FORTRAN main program, as all the development tools (such as symbolic debuggers and profilers) will be available. This may be the most effective way to trace 'memory leaks' (allocating but not freeing storage) and writing outside arrays. (Array-bound checking can often be used for FORTRAN code called from a FORTRAN main program. That can include R if compiled suitably.)

Our experience as teachers is that by far the most common errors are in the S / C interface, and there is no way to check consistency there automatically. So do check that the numbers and types of the arguments match exactly (including int *vs* long in C), and that the S objects are really known to have the correct type. (See page 203 for an example from S itself of what can go wrong if this check is not done.) Then check that the vector arguments passed to the compiled code are of sufficient length. (This can be particularly difficult to get right when using published code whose specified requirements are wrong! One way to check is to pass vectors that are longer than required, and check that the excess elements are unchanged by the call to the compiled code.)

## 6.6   Portability

Considerable care is needed to write code that will work on all platforms for one version of S-PLUS or R, let alone across engines. In the rest of this section we collect together the tips which we have found useful.

1. Include error-checking in your code. Surprisingly little user-contributed code has any.

2. Hide system-dependent definitions in header files such as verS.h on page 128.

3. Write in double precision. Not only does this aid porting to R, it is likely to minimize the surprises due to different rounding errors on different plat-forms. On perhaps the most widespread platform of all, Intel i386, double-precision operations are often as fast as single-precision ones.

4. If using mixed C and FORTRAN code do use the F77_XXX macros.

5. Note that an ANSI C compiler is not used on all S-PLUS UNIX platforms, and some of the compilers that are used reject ANSI constructs, especially prototypes. Our solution is to write and test in ANSI C but to distribute source code converted by the GNU utility unprotoize (part of gcc).

6. Check your code with tools such as lint and gcc -Wall and heed the warnings they give.

7. To avoid inadvertently overwriting existing symbols on some systems it is a good idea to declare as static all internal functions in C code. Then check that the symbols you do export do not cause conflicts. In S-PLUS on UNIX a formal check for potential conflicts can be made by

```
Splus NM myobj.o | cut -f 2-3 -d " " | grep "^T" \
    | cut -f 2 -d " " | sed 's/_$//' > t1
Splus NM 'Splus SHOME'/cmd/Sqpe | cut -f 2-3 -d " " \
    | grep "^T" | cut -f 2 -d " " | sed 's/_$//' > t2
comm -12 t1 t2
```

Note that we remove any trailing underscores from the symbol names, as
on HP systems symbol names from FORTRAN will not have trailing underscores.

However, no check will protect you from the writer of another library using
the same symbol, so the most reliable workaround is to use symbol names
that are likely to be unique to you. We use VR_* and BDR*, which was fine
until S-PLUS included some of our code without our knowledge, and used
our symbol names!

R has an argument PACKAGE that can be used with .C (and so on) to con-
fine the search for symbols to a single shared library or DLL.

8. Check as carefully as you can that storage allocated by Calloc will be
   freed. What happens if the user interrupts your code? (You may want an
   on.exit action.)

9. Remember that if you use NAOK or specialsok you need to consider NAs
   and/or specials in *all* of the arguments (even the integer ones).

# Chapter 7

# General Strategies and Extended Examples

Most S-PLUS programmers find early in their careers that they have written code which exhausts the physical memory (RAM) available to the S-PLUS process, and then proceeds to spend almost all its time in allocating virtual memory. Such code can reduce the fastest workstation to page thrashing, which will reduce severely the size of problem which can be tackled. This chapter give some hints on using S more efficiently.

The issues in R are related but different: R has a fixed workspace[1] which must be sized to fit into RAM or page thrashing will probably take most of the time, and evaluations that need more than the available workspace will fail. R

There are various tools available in S-PLUS to measure resources, but they differ between versions. S-PLUS 3.x and 5.x have memory.size, and S-PLUS 4.x and 5.x have mem.tally.reset and mem.tally.report. Further, S-PLUS 3.4 has the internal code for mem.tally.* but lacks the S wrappers.[2] Using these we can write a resources function[3] which works on 3.4 or later as

```
resources <- function(expr) {
  loc <- sys.parent(1)
  if(loc == 1) loc <- F
  on.exit(cat("Timing stopped at:", proc.time() - time, "\n"))
  expr <- substitute(expr)
  stime <- proc.time()
  mem.tally.reset()
  w <- eval(expr, local = loc)
  etime <- proc.time()
  mem <- mem.tally.report()
  on.exit()
  names(mem) <- c("Cache", "Working")
  # proc.time() gives length 1 on Windows.
```

---

[1] at least in the version (0.90.1) current when this was written.

[2] which we give in the scripts for this chapter.

[3] based with permission on code by Bill Dunlap of MathSoft DAPD.

151

```
if(length(stime) == 1) stime <- c(NA, NA, stime, 0, 0)
if(length(etime) == 1) etime <- c(NA, NA, etime, 0, 0)
time <- etime - stime
time[3] <- max(time[3], time[1] + time[2])
print(c(CPU = time[1] + time[2],
        Elapsed = time[3],
        "% CPU" = round((100 * (time[1] + time[2]))/time[3], 1),
        Child = time[4] + time[5], mem))
invisible(w)
}
```

This reports on two types of memory usage: cache memory used for keeping S objects in memory for faster access, and the maximum amount of memory used during the evaluation of `expr`. The use of cache can be controlled by the argument `keep` to `options`; by default only functions are cached.

Since our aim is not to compare systems, the timings here using different engines were done on different systems, all of which had ample RAM. It is worth noting that the different S implementations used here do differ, sometimes radically, in their ordering of approaches, and the ordering might be different again on machines with less RAM available.

Let us try `resources` on a simple example of bad S programming, to produce 1 000 sorted normal samples of size 100.

```
resources({
    X <- NULL
    for(i in 1:1000) X <- cbind(X, sort(rnorm(100)))
})
```

| CPU | Elapsed | % CPU | Child | Cache | Working | |
|---|---|---|---|---|---|---|
| | | | | | failed | (3.4) |
| | 15.47 | | | 9492 | 11 989 741 | (2000) |
| 10.47 | 11 | 95.2 | 0 | 7408 | 45 516 | (5.1) |
| 18.17 | 19 | | | | | (R) |

S-PLUS 3.4 ran out of the available memory (at 91Mb).

We can try some potential improvements

```
resources({
    X <- matrix(0, 100, 1000)
    for(i in 1:1000) X[, i] <- sort(rnorm(100))
})
```

| CPU | Elapsed | % CPU | Child | Cache | Working | |
|---|---|---|---|---|---|---|
| 2.39 | 3 | 79.7 | 0 | 15352 | 1 210 648 | (3.4) |
| | 1.32 | | 0 | 0 | 1 209 636 | (2000) |
| 2.26 | 3 | 75.3 | 0 | 0 | 828 896 | (5.1) |
| 1.35 | 2 | | | | | (R) |

```
resources(
    X <- apply(matrix(rnorm(100000), 100, 1000), 2, sort)
)
```

```
CPU Elapsed % CPU Child Cache     Working
3.77       4  94.3      0 72720  8 075 712 (3.4)
           1.98         0 63720  8 089 876 (2000)
2.38     2.38  100      0 71140  1 655 216 (5.1)
2.20     2.2                               (R)


resources({
    X <- rnorm(100000)
    X <- matrix(X[order(rep(1:1000, 100), X)], 100, 1000)
})
  CPU Elapsed % CPU Child Cache     Working
  2.4       3   80      0 17776  6 032 322 (3.4)
          1.66         0 18984  6 033 842 (2000)
 1.52       2   76      0 31792  2 939 616 (5.1)
 1.83       2                              (R)
```

The last version needs a little explanation. The generated normals are tagged by $1:1000$, 100 of each (the order does not matter) and then sorted by tag and then within groups with the same tag.

The S-PLUS function `memory.size` is provided to help trace where memory is being allocated. It returns the size of the currently allocated memory in bytes. The function `allocated` gives more detail of the memory usage.

Much of the material in this chapter is on efficiency issues. There is more to efficiency than CPU time and memory usage, and more resources that should be considered (not least programming effort). Some issues of memory usage that concerned us only a few years ago now seem quaint, because large amounts of RAM have become much more readily available and because the S implementations manage memory better. Nevertheless, memory issues do bite quite often. One example is the data-processing code used for the `Aids3` dataset in MASS (page 395 of third edition), a modestly large dataset with about 3000 rows. The code in the first two editions used a loop, and was run happily on a Sun workstation with 12Mb RAM in early 1993. By late 1998 it would not run in S-PLUS 5.0 in 64Mb, nor with R on a PC with 32Mb, and we changed to a partially vectorized solution that uses more memory on S-PLUS 3.4 but much less on the other two engines. On the other hand, running the scripts for the survival chapter in MASS took about 90 minutes in 1993 and now takes 5 minutes. Much of the time the issue is not whether the calculation is too slow but whether it can be completed in the resources available.

## 7.1 Managing loops

A major issue is that S is designed to be able to back out from uncompleted calculations, so that the memory used in intermediate calculations is retained until they are committed. This applies to `for`, `while` and `repeat` loops, for which none of the steps are committed until the whole loop is completed. (In recent versions of S-PLUS attempts are made to release unused memory at the end of each iteration, which has helped considerably.)

It is worth thinking hard if a loop is really necessary. It is rarely necessary to loop over the index of a vector or list: the computation should if at all possible be vectorized (the `ifelse` function is often handy) or a function such as `sapply` or `lapply` can be used. Operations on arrays can often be partially vectorized, and simple matrix operations can be replaced by elementwise or matrix multiplication.

Not all calculations can be vectorized, especially those that depend on the result of the previous calculation. The functions `cumsum`, `cumprod`, `cummax` and `cummin` are sometimes useful to vectorize calculations of this sort.

It was once true that functions such as `sapply` concealed explicit loops, but this is no longer the case in the S engines. Using standard constructions is likely to benefit most from future performance improvements in S-PLUS.

One tip is to make the whole body of a `for` loop a call to a function which returns nothing, or at least end it with `NULL`: S-PLUS 3.x retains the result of the last expression in each iteration until the end of the `for` loop, and generally the simpler the loop the better the code to reclaim memory will work.

## Distribution of a determinant

Consider the problem of finding the distribution of the determinant of a $2 \times 2$ matrix where the entries are independent and uniformly distributed digits $0, 1, \ldots, 9$. This amounts to finding all possible values of $ac - bd$ where $a$, $b$, $c$ and $d$ are digits. One solution we were shown is

```
resources({
  val <- NULL
  for(a in 0:9)
    for(b in 0:9)
      for(d in 0:9)
        for(e in 0:9)
          val <- c(val, a*b - d*e)
  freq <- table(val)
})
```

| CPU | Elapsed | % CPU | Child | Cache | Working | |
|---|---|---|---|---|---|---|
| 10.3 | 10.3 | 100 | 0 | 46864 | 2 122 568 | (3.4) |
| | 3.2 | | 0 | 32520 | 848 698 | (2000) |
| 6.02 | 6.02 | 100 | 0 | 640 | 14 900 | (5.1) |
| 6.35 | 7.00 | | | | | (R) |

The 'obvious' choice of a to d as loop indices would have given warnings on some systems, and wrong answers or even an infinite loop on some.

A much faster way is to use `outer`:

```
resources({
  val <- outer(0:9, 0:9, "*")
  val <- outer(val, val, "-")
  freq <- table(val)
```

```
})
  CPU Elapsed % CPU Child Cache    Working
  0.16    0.16   100     0 26664   437 713 (3.4)
          0.09           0 42012   556 509 (2000)
  0.11    0.11   100     0 33652   173 300 (5.1)
  0.23    0.23                              (R)
```

Notice that the third argument to `outer` may be a character string giving the name of a function of two arguments. We can speed this up a bit more by using the function `tabulate` directly rather than `table`.

```
resources({
    val <- outer(0:9, 0:9, "*")
    val <- outer(val, val, "-")
    freq <- tabulate(val + 82)
})
  CPU Elapsed % CPU Child Cache    Working
  0.12    0.12   100     0 48480   258 712 (3.4)
          0.03           0     0   372 556 (2000)
  0.06    0.06   100     0     0   173 300 (5.1)
  0.02    0.02                              (R)
```

## Simulation of a correlation coefficient

Loops occur quite often in simulation problems. Here is a real-life one from a posting to the `S-news` discussion list, which occurs in connection with the Shapiro-Wilk test for normality.

'Simulate a sorted sample of size 10 000 from the distribution of $1 - r^2$ where $r$ is the correlation coefficient between a sorted normal sample of size 10 and fixed vector $a$.'

One naive approach would be

```
a <- 1:10
resources({
    tmp <- matrix(10, 10000, 1)
    tmp <- t(apply(tmp, 1, rnorm))
    tmp <- t(apply(tmp, 1, sort))
    SWstat <- function(y) {return(1 - cor(y, a)^2)}
    tmp <- apply(tmp, 1, SWstat)
    tmp <- sort(tmp)
})
   CPU Elapsed % CPU Child    Cache        Working
   67.7      70  96.8     0   137360   12 134 784 (3.4)
          130.84          0   159960   15 000 772 (2000)
    610     617    99     0  5521884   14 195 245 (5.1)
  26.07      27                                   (R)
```

If $y$ is one such sorted normal sample then the correlation coefficient is

$$r = \sum y_i^\star a_i^\star \qquad \text{where} \qquad y_i^\star = (y_i - \bar{y})/\sqrt{\sum (y_i - \bar{y})^2}$$

and $a_i^\star$ is defined similarly. We can use this to reduce the calculations.

```
resources({
    a <- a - mean(a)
    a <- a/sqrt(sum(a^2))
    tmp <- rnorm(10*10000)
    tmp <- matrix(tmp[order(rep(1:10000, 10), tmp)], 10, 10000)
    mns <- rep(1,10) %*% tmp / 10
    ten <- rep(10, 10000)
    tmp <- tmp - rep(mns, ten)
    cln <- sqrt(rep(1,10) %*% (tmp * tmp))
    tmp <- tmp/rep(cln, ten)
    rm(mns, cln)
    tmp <- sort(1 - (a %*% tmp)^2)
})
```

| CPU | Elapsed | % CPU | Child | Cache | Working | |
|-----|---------|-------|-------|-------|---------|--|
| 3.54 | 4 | 88.5 | 0 | 64640 | 6 672 320 | (3.4) |
| | 2.36 | | 0 | 67764 | 6 076 566 | (2000) |
| 2.25 | 2.25 | 100 | 0 | 107640 | 2 876 280 | (5.1) |
| 1.65 | 1.65 | | | | | (R) |

We could make it look more streamlined by using scale, but there may be a penalty in both time and memory usage.

```
resources({
    tmp <- rnorm(10*10000)
    tmp <- scale(matrix(tmp[
        order(rep(1:10000, 10), tmp)], 10, 10000))
    tmp <- sort(1 - (t(scale(a)) %*% tmp / 9)^2)
})
```

| CPU | Elapsed | % CPU | Child | Cache | Working | |
|-----|---------|-------|-------|-------|---------|--|
| 3.65 | 4 | 91.3 | 0 | 12928 | 7 936 968 | (3.4) |
| | 2.25 | | 0 | 51504 | 7 082 724 | (2000) |
| 2.16 | 2.16 | 100 | 0 | 90156 | 2 895 724 | (5.1) |
| 12.69 | 13 | | | | | (R) |

There is often some advantage in splitting a vectorized calculation into several sections and looping over the sections.

```
resources({  verbose <- T
    a <- a - mean(a)
    a <- a/sqrt(sum(a^2))
    sws <- numeric(10000)
    ind <- rep(1:1000, 10)
    ten <- rep(10, 1000)
    u <- rep(1,10)
    for(i in 0:9) {
```

```
        tmp <- rnorm(10*1000)
        tmp <- matrix(tmp[order(ind, tmp)], 10, 1000)
        tmp <- tmp - rep(u %*% tmp / 10, ten)
        tmp <- tmp/rep(sqrt(u %*% (tmp * tmp)),ten)
        sws[1000*i + 1:1000] <- 1 - (a %*% tmp)^2
        if(verbose) cat("Slab", i + 1, "finished.\n")
    }
    rm(tmp, ind, ten, u)
    sws <- sort(sws)
})
  CPU Elapsed % CPU Child Cache   Working
  3.22    3.22   100     0     0  791 753 (3.4)
          1.74           0     0  759 683 (2000)
  1.91       2  95.5     0  3368  479 772 (5.1)
  1.21    1.21                         (R)
```

## For loops

S-PLUS also has For loops, which are similar to `for` loops but 'unroll' the loop and (by default) use a separate S-PLUS process to perform the calculations. This may be beneficial in reducing memory usage. For is a function, so we translate

```
res <- numeric(10)
for(i in 1:10) res[i] <- myfunction(x[i,])
```

to

```
res <- numeric(10)
For(i = 1:10, res[i] <- myfunction(x[i,]), debug = T)
```

Using the argument `debug = T` will show the script that will be run, here

```
options(error = function(){dump.calls(); q(1)})
assign('.Program', w=0, expression(eval(parse())))
assign('.Steps', .Steps, w=0) ; remove('.Steps', w=1)
{i <- .Steps[[1]]; res[i] <- myfunction(x[i, ])}
    ....
{i <- .Steps[[10]]; res[i] <- myfunction(x[i, ])}
```

As this will be run in a separate process, it may be necessary to specify initialization action in the `first` argument, for example calls to `library`.

The argument `grain.size` affects the script which is written. The default, 1, gives the script already shown in which each step of the loop is a separate top-level expression enclosed in braces. If we use

```
res <- numeric(10)
For(i = 1:10, res[i] <- myfunction(x[i,]), debug=T, grain.size=5)
```

we get

```
options(error = function(){dump.calls(); q(1)})
assign('.Program', w=0, expression(eval(parse())))
assign('.Steps', .Steps, w=0) ; remove('.Steps', w=1)
{i <- .Steps[[1]]; res[i] <- myfunction(x[i,  ])
   ....
 i <- .Steps[[5]]; res[i] <- myfunction(x[i,  ])
}
{i <- .Steps[[6]]; res[i] <- myfunction(x[i,  ])
   ....
 i <- .Steps[[10]]; res[i] <- myfunction(x[i,  ])
}
```

grouping `grain.size` steps into one top-level expression. This will reduce the frequency of commitment to the working directory, thereby reducing I/O and increasing memory usage.

We can try this on our example.

```
resources({  verbose <- T
  a <- a - mean(a)
  a <- a/sqrt(sum(a^2))
  sws <- numeric(10000)
  ind <- rep(1:1000, 10)
  ten <- rep(10, 1000)
  u <- rep(1,10)
  For(i = 0:9, {
     tmp <- rnorm(10*1000)
     tmp <- matrix(tmp[order(ind, tmp)], 10, 1000)
     tmp <- tmp - rep(u %*% tmp / 10, ten)
     tmp <- tmp/rep(sqrt(u %*% (tmp * tmp)),ten)
     sws[1000*i + 1:1000] <- 1 - (a %*% tmp)^2
     if(verbose) cat("Slab", i + 1, "finished.\n")
  }, exec=T)
  rm(ind, ten, u)
  sws <- sort(sws)
})
```

| CPU | Elapsed | % CPU | Child | Cache | Working | | |
|------|---------|-------|-------|--------|------|-----|-------|
| 0.10 | 10 | 1 | 3.77 | 52520 | 429 | 608 | (3.4) |
|  | 7.02 |  |  | 48784 | 316 | 348 | (2000) |
| 0.17 | 3 | 5.7 | 2.51 | 187440 | 340 | 172 | (5.1) |

```
resources({  verbose <- T
  a <- a - mean(a)
  a <- a/sqrt(sum(a^2))
  sws <- numeric(10000)
  ind <- rep(1:1000, 10)
  ten <- rep(10, 1000)
  u <- rep(1,10)
  For(i <- 0:9, {
     tmp <- rnorm(10*1000)
     tmp <- matrix(tmp[order(ind, tmp)], 10, 1000)
```

```
        tmp <- tmp - rep(u %*% tmp / 10, ten)
        tmp <- tmp/rep(sqrt(u %*% (tmp * tmp)),ten)
        sws[1000*i + 1:1000] <- 1 - (a %*% tmp)^2
        if(verbose) cat("Slab", i + 1, "finished.\n")
      }, grain.size = 5, exec=T)
      rm(ind, ten, u)
      sws <- sort(sws)
   })
```

| CPU | Elapsed | % CPU | Child | Cache | Working | |
|---|---|---|---|---|---|---|
| 0.08 | 5 | 1.6 | 3.72 | 21008 | 388 968 | (3.4) |
| | 4.54 | | | 29800 | 312 304 | (2000) |
| 0.14 | 3 | 4.7 | 2.44 | 37332 | 217 580 | (5.1) |

Note that while the elapsed timings are informative, the memory usage refers
only to the main S-PLUS process. The second S-PLUS process is taking up to
another 10M of RAM not accounted for here. For loops are most useful when
the memory usage is large. They are not implemented in R, and are only useful   R
on S-PLUS 4.5 and 2000 amongst current Windows versions of S-PLUS, as on
those it runs the lightweight Sqpe.exe as the child process.


## 7.2   A large regression

The following is based on a real example posted to S-news. The poster wanted
to do a regression of 17 065 cases with several (say four) continuous variables and
one factor with 107 levels. Her machine did not have enough memory to use lm
to do this, as the design matrix is $17065 \times 111$, that is 15Mb in size, although it
can be done, slowly, on a machine with enough RAM.

Both of us responded with solutions based on the special structure of the problem. Solution Lreg1 is based on using the modified Gram-Schmidt process for
orthogonalization, and solution Lreg2 on constructing $X'X$ and $X'y$ without
constructing $X$. After some polishing we can test them out:

```
Lreg1 <- function(fac, X0, y) {
  mc <- function(x, f)    # 'mean correct'
    x - (unlist(lapply(split'(x, f), mean)))[f]
  rc <- function(X, y)    # 'regression coef'
    solve(crossprod(X), crossprod(X, y))
  b2 <- rc(apply(X0, 2, mc, fac), mc(y, fac))
  b1 <- unlist(lapply(split(y - X0%*%b2, fac), mean))
  c(b1, b2)
}

Lreg2 <- function(fac, X, y) {
  ni <- tabulate(unclass(fac))
  p1 <- length(ni); p <- ncol(X)
  XX <- matrix(0, p1 + p, p1 + p)
  diag(XX) <- c(ni, rep(0, p))
```

```
    XX[p1 + (1:p), p1 + (1:p)] <- crossprod(X)
    XX[p1 + (1:p), 1:p1] <-
      t(XX[1:p1, p1 + (1:p)] <-
        matrix(apply(X, 2, function(x, f)
                        unlist(lapply(split(x, f), sum)), fac),
              p1, p))
    b <- c(unlist(lapply(split(y, fac), sum)), crossprod(X, y))
    drop(solve(XX, b))
}


N <- 17065
fac <- factor(sample(1:107, N, rep=T))
X0 <- matrix(rnorm(4*N), ncol=4)
y <- (1:107)[fac] + X0 %*% rep(1, 4) + rnorm(N, 0, 0.25)

options(object.size=2e7) # allow 20Mb objects
options(contrasts=c("contr.treatment", "contr.poly"))
resources(v0 <- lm(y ~ fac + X0 - 1))
   CPU Elapsed % CPU Child  Cache      Working
  28.9      59    49     0  66256  54 029 136 (3.4)
          17.18          0  60996  54 815 596 (2000)
   8.9      10    89     0 254744  30 027 720 (5.1)
  15.0      16                       ca 80Mb  (R)

resources(v1 <- Lreg1(fac, X0, y))
   CPU Elapsed % CPU Child  Cache     Working
  1.81       3  60.3     0 170488  5 912 296 (3.4)
          1.04          0 155916  5 911 256 (2000)
  2.02    2.02   100     0 140104  1 041 312 (5.1)
  1.17    1.17                               (R)

resources(v2 <- Lreg2(fac, X0, y))
   CPU Elapsed % CPU Child  Cache     Working
  1.13    1.13   100     0  39592  3 531 952 (3.4)
           0.8          0  23028  3 168 180 (2000)
  0.63       1    63     0  35664    276 176 (5.1)
  0.45       1                               (R)
```

We can fairly easily extend Lreg2 to find the variance matrix of the coefficients
and the residuals

```
    ....
    b <- c(unlist(lapply(split(y, fac), sum)), crossprod(X, y))
    b <- drop(solve(XX, b))
    resid <- y - (b[fac] + X %*% b[p1 + (1:p)])
    sigma2 <- sum(resid^2)/(length(y) - p - p1)
    list(coef = b, resid = resid, var = sigma2*solve(XX))
}
```

and anything else we might want.

## 7.3   Simulation envelopes for normal-scores plots

Simulation envelopes are a graphical model-checking device for linear models, along the lines of the example in MASS page 72 applied to the standardized residuals. (Atkinson, 1985, has many similar plots, as half-normal plots for the absolute values.)

1. Fit the linear model, calculate the residuals and standardise them to unit variance.

2. Generate $N$ ($\approx 1000$) normal samples as responses. For each sample fit the same linear model and calculate the standardized residuals.

3. Sort all of the sets of standardized residuals.

4. The envelope consists of the lower and upper 2.5% quantiles of the $N$ generated standardized residuals at each ordered position.

   Here is one way (based closely on code given by Everitt, 1994, p. 37) that we do not recommend.

```
env1 <- function(obj) {
  res <- resid(obj); n <- length(res)
  H <- model.matrix(obj); p <- ncol(H)
  res <- res/sqrt(sum(res^2)/(n-p))
  H <- H %*% solve(t(H) %*% H)  %*% t(H)
  ident <- diag(n)
  epsilon <- matrix(0, n, 1000)
  e <- matrix(0, n, 1000)
  e1 <- numeric(n); e2 <- numeric(n)
  for(i in 1:1000) {
    epsilon[, i] <- rnorm(n, 0, 1)
    e[, i] <- (ident - H) %*% epsilon[, i]
    e[, i] <- sort(e[, i])/sqrt(sum(e[, i]^2)/(n-p))
  }
  for(i in 1:n) {
    eo <- quantile(e[i, ], c(0.025, 0.975))
    e1[i] <- eo[1]; e2[i] <- eo[2]
  }
  ylim <- range(res, e1, e2)
  oldpar <- par(pty = "s"); on.exit(par(oldpar))
  qqnorm(res, pch = 1, ylim = ylim, ylab = "Std. Residuals",
         main = deparse(obj$call$formula))
  par(new = T)
  qqnorm(e1, axes=F, xlab="", ylab="", type="l", ylim=ylim)
  par(new = T)
  qqnorm(e2, axes=F, xlab="", ylab="", type="l", ylim=ylim)
  invisible(obj)
}
```

Our solution vectorizes the problem and makes uses of classes to separate computation from plotting and printing.

```
env2 <- function(object, nsamples = 1000, alpha = 0.05) {
  n <- length(r <- resid(object))
  p <- length(coef(object))
  Y <- scale(qr.resid(qr(model.matrix(object)),
        matrix(rnorm(n * nsamples), n, nsamples)),
                    F, T) * sqrt((n - 1)/(n - p))
  Y[, ] <- Y[order(col(Y), Y)]
  Y <- matrix(Y[order(row(Y), Y)], nsamples, n)
  x0 <- quantile(1:nsamples, c(alpha/2, 1-alpha/2))
  if(all(x0 %% 1 == 0)) elim <- t(Y[x0, ])
  else {
    x1 <- c(floor(x0), ceiling(x0))
    elim <- cbind(
        Y[x1[1], ] + (Y[x1[3], ] - Y[x1[1], ])/
        (x1[3] - x1[1])*(x0[1] - x1[1]),
        Y[x1[2], ] + (Y[x1[4], ] - Y[x1[2], ])/
        (x1[4] - x1[2])*(x0[2] - x1[2]))
  }
  res <- sort(r)
  res <- res/sqrt(sum(res^2)/(n-p))
  res <- structure(list(res=res, elim=elim),
              label = deparse(object$call$formula))
  class(res) <- "envelope" # or oldClass
  res
}

plot.envelope <- function(x, ...) {
  n <- length(x$res)
  ylim <- range(x$res[c(1,n)], x$elim[c(1,n),])
  nscores <- qnorm(ppoints(n))
  oldpar <- par(pty = "s"); on.exit(par(oldpar))
  plot(nscores, x$res, pch = 1, ylim = ylim,
      xlab = "Normal scores", ylab = "Sorted residuals",
      main = attr(x, "label"))
  lines(nscores, x$elim[,1]); lines(nscores, x$elim[,2])
  invisible(x)
}

print.envelope <- function(x, ...) {
  lo <- x$res < x$elim[,1]; hi <- x$res > x$elim[,2]
  flash <- rep("", length(lo))
  flash[lo] <- "<"; flash[hi] <- ">"
  if(any(lo | hi))
    print(cbind(do.call("data.frame",x), flash=flash)[lo | hi,])
  else cat("All points within envelope\n")
  invisible(x)
}
```

We can try this out on the data on Black Cherry trees: see Figure 7.1.

```
library(MASS)
```

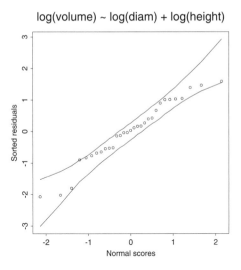

**Figure 7.1**: A simulation envelope for a normal-scores residual plot.

```
fm <- lm(log(volume) ~ log(diam) + log(height), trees)
resources(env1(fm))
  CPU Elapsed % CPU Child  Cache     Working
11.59      12  96.6     0  15352  2 598 976 (3.4)
         5.76           0 325368    736 928 (2000)
18.52      21  88.2  0.08 743844    371 867 (5.1)
 2.11       2           .                    (R)

print(resources(plot(env2(fm))))
  CPU Elapsed % CPU Child  Cache     Working
 1.77       2  88.5     0  73528  3 277 538 (3.4)
         0.82     0     0 100368  3 279 970 (2000)
 1.11    1.11   100     0  16972    738 683 (5.1)
 1.41       2                                (R)
```

Notice how the relative speed differs markedly by engine.

## 7.4   Making good use of language objects

In Section 3.5 we saw some of the power and flexibility that follows from the ability the S language has to construct and manipulate objects that are effectively the code that drives the engine itself, that is, using S to program in S. We now consider two slightly longer examples that illustrate some of the common idioms.

### Expectations of order statistics

If $X$ is a random variable with distribution function $F(x)$ and density $f(x)$ then the $r$ th order statistic from a sample of size $n$ has distribution function (Cox &

Hinkley, 1974, §A.2)

$$F_{n,r}(x) = \sum_{k=r}^{n} \binom{n}{k} \{F(x)\}^k \{1 - F(x)\}^{n-k}$$

and hence its density function is

$$f_{n,r}(x) = \frac{n!}{(r-1)!(n-r)!} \{F(x)\}^{r-1} \{1 - F(x)\}^{n-r} f(x)$$

The current S-PLUS help page for the `call` function gives one way to write an S function[4] to construct this density function as

```
d.order <- function(n, r, distfun, densityfun) { # S version
    f <- function(x) NULL
    con <- exp(lgamma(n + 1) - lgamma(r) - lgamma(n - r + 1))
    c1 <- call(substitute(distfun), as.name("x"))
    c2 <- call("^", call("-", 1, c1), n - r)
    c3 <- call("^", c1, r - 1)
    c4 <- call("*", con, call("*", c2, call("*", c3,
            call(substitute(densityfun), as.name("x")))))
    f[[length(f)]] <- c4
    f
}
```

An example is

```
> d.order(10, 4, pnorm, dnorm)
function(x)
840 * ((1 - pnorm(x))^6 * (pnorm(x)^3 * dnorm(x)))
```

This illustrates well how the function `call` can be used for this kind of S construction, but this function may be criticized on a number of grounds, including

(a) The distribution function, here `pnorm`, is evaluated twice with the same arguments,

(b) No provision is made for additional arguments to the distribution and density functions (for example the mean and standard deviation for a normal distribution).

(c) It is practically unreadable.

Let us consider a slightly expanded version of this problem. Suppose we are investigating approximations to the expectations of order statistics from various distributions and for reference we need values by direct numerical integration of

$$E X^{(r)} = \int_{-\infty}^{\infty} x f_{n,r}(x) \, dx$$

We will provide two functions, one to construct the integrand and another to use `integrate` to find the expectations. One way of writing the first is

---

[4]it needs a few changes for R given in the on-line scripts for this chapter.

```
make.integrand <- function(n, r, cdf, pdf) {
   cdf <- substitute(cdf)
   if(is.name(cdf)) cdf <- call(cdf, as.name("x"))
   pdf <- substitute(pdf)
   if(is.name(pdf)) pdf <- call(pdf, as.name("x"))
   con <- round(exp(lgamma(n+1) - lgamma(r) - lgamma(n-r+1)))
   substitute(function(x) {
               P <- cdf
               K * x * P^r1 * (1 - P)^nr * pdf
               }, list(K = con, cdf = cdf, r1 = r - 1,
                        nr = n - r, pdf = pdf))
}
```

The idea is to use substitute first to capture the arguments and later to make genuine substitutions in a template function. The user may provide either names or calls for the distribution and density function arguments, and additional arguments may be provided by giving the full call. Two examples are

```
> make.integrand(10, 4, pnorm, dnorm)
function(x)
{
        P <- pnorm(x)
        840 * x * P^3 * (1 - P)^6 * dnorm(x)
}
> make.integrand(10, 4, pt(x, 5), dt(x, 5))
function(x)
{
        P <- pt(x, 5)
        840 * x * P^3 * (1 - P)^6 * dt(x, 5)
}
```

An R version of this function can make use of the enclosing environment (rather      R than substitution) to achieve the same effect

```
make.integrand <- function (n, r, cdf, pdf) {     # R only
   cdf <- substitute(cdf)
   if (is.name(cdf)) cdf <- call(deparse(cdf), as.name("x"))
   pdf <- substitute(pdf)
   if (is.name(pdf)) pdf <- call(deparse(pdf), as.name("x"))
   con <- round(exp(lgamma(n+1) - lgamma(r) - lgamma(n-r+1)))
   function(x) {
     P <- eval(cdf); D <- eval(pdf)
     x * con * P^(r - 1) * (1 - P)^(n - r) * D
   }
}
```

Notice that in R the first argument to call must be a character string but in S it may be a name or a character string. We can devise a much simpler solution for R (see below) if we do not require it to perform in precisely the same way as the present version.

Now consider a function to compute the expectations.

```
E.order <- function(n, cdf, pdf, lower = -Inf, upper = Inf, ...)
{
  Ex <- numeric(n)
  for(i in 1:n) {
    f <- eval(call("make.integrand", n, i,
                   substitute(cdf), substitute(pdf)))
    Ex[i] <- integrate(f, lower, upper, ...)[[1]]
  }
  Ex
}
```

Notice that to transmit the arguments to the outer function verbatim we need to construct the call to `make.integrand` and evaluate it, effectively the same technique as `do.call` uses. Allowing the user to provide either a name or a call for the distribution and density function arguments has the mild disadvantage of requiring any additional arguments to be specified twice, but has the compensating advantages of hard-wiring them into the integrand and of leaving the ... argument to the outer function free and available for transmission to `integrate`.[5]

We can now compare using our function with two commonly used approximations for a specific normal case.

```
> qnorm((1:10 - 1/2)/10)
 [1] -1.64485 -1.03643 -0.67449 -0.38532 -0.12566  0.12566
 [7]  0.38532  0.67449  1.03643  1.64485
> qnorm((1:10 - 3/8)/10.25)
 [1] -1.54664 -1.00049 -0.65542 -0.37546 -0.12258  0.12258
 [7]  0.37546  0.65542  1.00049  1.54664
> E.order(10, pnorm, dnorm)
 [1] -1.53875 -1.00136 -0.65606 -0.37576 -0.12267  0.12267
 [7]  0.37576  0.65606  1.00136  1.53875
```

*Another approach*

As with most S problems there are many possible solutions. In yet another approach we pass `cdf` and `pdf` as default arguments. The following version for the S engines[6] allows us to specify common extra arguments for the pdf and cdf,

```
make.integrand <- function(n, r, cdf, pdf, ...) {
  con <- round(exp(lgamma(n+1) - lgamma(r) - lgamma(n-r+1)))
  f <- substitute(function(x, cdf, pdf) {
                  P <- cdf(x)
                  K * x * P^r1 * (1 - P)^nr * pdf(x)
                  }, list(K = con, r1 = r - 1, nr = n - r))
  dots <- list(...)
  if(length(dots)) {
    cdf[names(dots)] <- unlist(dots)
```

---

[5]At the time of writing the `integrate` function in R has different names for the components of its values from those of S, and did not handle infinite ranges.

[6]for use with R we need to use `formals` as in `formals(cdf)[names(dots)] <- unlist(dots)`, and to `eval` the substitution.

```
      pdf[names(dots)] <- unlist(dots)
    }
    f$cdf <- cdf; f$pdf <- pdf
    f
  }
  E.order <- function(n, cdf, pdf, lower = -Inf, upper = Inf, ...)
  {
    Ex <- numeric(n)
    for(i in 1:n) {
      f <- make.integrand(n, i, cdf, pdf, ...)
      Ex[i] <- integrate(f, lower = lower, upper = upper)[[1]]
    }
    Ex
  }
```

used by, for example `E.order(10, pt, dt, df=5)`.

An alert reader might wonder why we do not pass the `...` argument to `integrate` and let that pass the extra arguments on to `make.integrand`. Unfortunately, the way `integrate` is implemented in the S engines only named arguments of the integrand can be passed down. However, this restriction does not apply to R, so `make.integrand` can be as simple as                                         R

```
  make.integrand <- function(n, r, cdf, pdf) {     # R only
    con <- round(exp(lgamma(n+1) - lgamma(r) - lgamma(n-r+1)))
    function(x, ...) {
      P <- cdf(x, ...)
      x * con * P^(r-1) * (1-P)^(n-r) * pdf(x, ...)
    }
  }
```

(based on an idea of Luke Tierney), used with

```
  E.order <- function(n, cdf, pdf, lower = -10, upper = 10, ...)
  {
    Ex <- numeric(n)
    for(i in 1:n) {
      f <- make.integrand(n, i, cdf, pdf)
      Ex[i] <- integrate(f, lower = lower, upper = upper,
                         ...)[[1]]
    }
    Ex
  }
```

Often using a combination of approaches like this pays off in a simple, comprehensible solution.

## Transformations of parameters

A common problem in practical statistics is to estimate a non-linear function of the parameters estimated from a fitted model. For example for the tree volume data used above for the display in Figure 7.1 on page 163 a natural quantity to estimate is the volume of the tree rather than the log-volume. With an obvious notation the *median* estimator is

$$\widehat{V} = \exp(\widehat{\beta_0} + \widehat{\beta_1}\log D + \widehat{\beta_2}\log H)$$

The estimate itself may be calculated by substitution, but in statistics no estimate is complete without some estimate of its uncertainty, at least an approximate one.

The delta method is the usual first step in finding such a rough estimate of error. It is obtained by supposing that the non-linear function can be adequately approximated by the tangential linear approximation at the estimate. That is, if $\theta = \theta(\beta)$ is a function of parameters $\beta$ then small variations in $\theta$ are approximately linearly related to small variations in the components of $\beta$ by

$$\delta\theta \approx \sum \frac{\partial\theta}{\partial\beta_j}\delta\beta_j$$

a notation that gives the method its name.

The parameters of linear models in S are labelled by the associated column name in the model matrix. Here we do need to have them named conveniently and for simplicity we will adopt the convention that an intercept parameter will be named b0 and others b1, b2, and so on, in the order they occur in the model matrix. Our first task is to write a generic suite of functions to extract and assemble estimate and variance information from a fitted model. We can use

```
estimate <- function(object, ...)
  UseMethod("estimate", object, ...)

estimate.lm <- function(object, ...) {
  b <- coef(object)
  if(m <- match("(Intercept)", names(b), nomatch = 0)) {
    l <- length(b) - 1
    if(l > 0) names(b)[-m] <- paste("b", 1:l, sep = "")
    names(b)[m] <- "b0"
  } else {
    names(b) <- paste("b", seq(along = b), sep = "")
  }
  V <- vcov(object)
  dimnames(V) <- list(names(b), names(b))
  estimate <- list(parameter = b, variance = V)
  class(estimate) <- "estimate"
  estimate
}

estimate.nls <- function(object, ...)
  structure(list(parameter = coef(object),
                 variance = vcov(object)), class = "estimate")

as.estimate <- function(x)
```

```
    if(inherits(x, "estimate")) x else estimate(x)

    is.estimate <- function(x) inherits(x, "estimate")

    print.estimate <- function(x, ...) {
      print(cbind(Parameters = x$parameter,
                  SE = sqrt(diag(x$variance))))
      invisible(x)
    }
```

Since `aov` and `glm` both inherit from `lm` no separate methods are needed for them, but other methods could usefully be added, such as for the models of parametric survival analysis.

Objects of class `"estimate"` are guaranteed to have a vector of parameters with names and a variance matrix with dimnames. We now consider writing a method for `estimate` for objects of class `"estimate"`. This function, `estimate.estimate`, will require, in addition to the object, a specification of the transformations required and a `data` argument to supply additional variables, if needed.

We will use the function `D` to differentiate the transformations symbolically. This makes the function very convenient, but somewhat limits the scope of transformations that it can accommodate. It requires the transformations to be supplied to the function as `call` objects, for which a convenient form is a formula. (If there are several functions this argument will be a list of formulae.) The left hand side of the formula will be used as a name for the transform but it is needed only if further transformations are needed. The right hand side specifies the transformation itself as a function of the parameters and possibly of additional variables.

In some cases only the standard errors of the results are needed. This option will be specified by an additional argument `se.only = T`, when the method function will return a data frame with two columns giving the estimates and standard errors.

In working through the method function shown in Figure 7.2 it may help to keep in mind the following:

(a) If there is a list of more than one transformation the values will initially be a list of vector values concatenated by `unlist` into one numeric vector. These components are generated separately by the internal function `getval`.

(b) Function `as.list` in the old S engine loses the names of a vector, whereas the other two engines use the names to name the list components. We can add the names back with a call to `names<-`, but then under the new S engine `pframe` has class `"named"` and `substitute` does not accept objects of mode `list` and class `"named"`. Compatibility!

(c) Notice that evaluating the transformations requires parameter estimates from the primary object and possibly additional variables from the `data` argument. However `eval` only allows one `local` argument for supplied variables. We could combine the `data` and parameters into one list to be supplied to `eval`

```
estimate.estimate <-
  function(object, transforms, data = sys.parent(),
         se.only = F, ...)
{
  getval <- function(x, pframe, data)
    eval(do.call("substitute", list(x[[length(x)]], pframe)),
        data)

  getblock <- function(transform, pframe, data) {
    getcol <- function(x, fun, pframe, data)
      eval(do.call("substitute", list(D(fun, x), pframe)),
          data)

    fun <- transform[[length(transform)]]
    do.call("cbind", lapply(names(pframe), getcol,
        fun = fun, pframe = pframe, data = data))
  }

  pframe <- as.list(object$parameter)
# needed in old S engine, only
# names(pframe) <- names(object$parameter)
  if(inherits(transforms, "formula"))
    transforms <- list(transforms)
  tnames <- unlist(lapply(transforms, function(x)
                      if(length(x) == 3) x[[2]] else ""))
  tvalues <-
    lapply(transforms, getval, pframe = pframe, data = data)
  names(tvalues) <- tnames
  tvalues <- unlist(tvalues)

  X <- do.call("rbind", lapply(transforms, getblock,
                      pframe = pframe, data = data))
  XV <- X %*% object$variance
  if(se.only) {
    se <- as.vector(sqrt((XV * X) %*% rep(1, ncol(X))))
    data.frame(Parameters = tvalues, SE = se)
  } else {
    structure(list(parameter = tvalues,
              variance = XV %*% t(X)),
          class = "estimate")
  }
}
```

**Figure 7.2**: A non-linear transformation function for estimates.

but this is difficult if `data` is not an explicit data frame but a frame number, as is the case with its default value.

The solution we have adopted is to substitute numbers for the parameters in the call object first and then to evaluate it with the `data` argument supplying any additional variables. This itself raises another problem, though. The function `substitute` does not evaluate its first argument but uses it verbatim as it is supplied. This means we have to construct the call to `substitute` itself so that the evaluator sees an explicit copy of the transformation and not a variable representing it. Hence the use of `do.call` within the `getval` and `getcol` internal functions.

(d) The variance matrix is calculated as a matrix product of the form $XVX^T$ where $X$ is the matrix of partial derivatives. It will have as many columns as there are original parameters and as many rows as there are new ones.

This matrix is put together with calls to `cbind` and `rbind`, but rather than work progressively with repeated calls to these functions the components are initially held in lists and a single call to each of `cbind` and `rbind` is made using the `do.call` function again. The main reason is to allow `cbind` and `rbind` to see all components at once so that short components can be detected and recycled as necessary.

For linear and non-linear models this function can be used as a non-linear version of `predict`. The `trees` example used to motivate this section may be handled by

```
> library(MASS)
> tree.lm <- lm(log(volume) ~ log(diam) + log(height), trees)
> tree.est <- estimate(tree.lm)
> tree.est
     Parameters       SE
b0      -6.6316 0.799790
b1       1.9826 0.075011
b2       1.1171 0.204437
> Trees <- expand.grid(diam=5*2:4, height=10*6:8)
> vol.est <- estimate(tree.est,
      ~exp(b0 + b1*log(diam) + b2*log(height)), Trees,
      se.only = T)
> m <- rbind(exp(predict(tree.lm, Trees)), vol.est$SE)
> dimnames(m) <- list(c("vol", "se"),
      apply(Trees, 1, function(x) paste(x, collapse=",")))
> round(m, 2)
        10,60 15,60 20,60 10,70 15,70 20,70 10,80 15,80 20,80
vol 12.27 27.42 48.51 14.58 32.58 57.63 16.93 37.82 66.90
 se  0.53  1.55  3.51  0.33  0.92  2.65  0.52  0.69  2.15
```

For a different example consider the `stormer` data used in MASS pp. 253–8. This is a non-linear regression of the form

$$t = \frac{\beta v}{w - \theta} + \varepsilon$$

that can also be fitted as a generalized linear model in the form

$$t = \frac{1}{\gamma_1 z_1 + \gamma_2 z_2} + \varepsilon$$

where $z_1 = w/v$, $z_2 = -1/v$, $\gamma_1 = 1/\beta$ and $\gamma_2 = \theta/\beta$. It is instructive to fit the model in one form and check by transformation that it agrees with the results when fitted in the other.

```
> storm.nls <- nls(Time ~ beta*Viscosity/(Wt - theta),
        stormer, start = c(beta = 29, theta = 2))
> estimate(storm.nls)
        Parameters        SE
 beta      29.4013 0.91553
 theta      2.2182 0.66552


> storm.glm <- glm(Time ~ I(Wt/Viscosity)+I(-1/Viscosity)-1,
        quasi(link = inverse, variance = constant), stormer,
        eps = 1.0e-8)
> storm.glm.est <- estimate(storm.glm)
> estimate(storm.glm.est, list(beta ~ 1/b1, theta ~ b2/b1))
        Parameters        SE
 beta      29.4013 0.91553
 theta      2.2183 0.66552
```

This shows good agreement.

## 7.5  Bootstrapping and cross-validation

The resources consumed by naive users of bootstrapping have been causing headaches for computing managers since the early days of the technique, and they are still a frequent topic of discussion amongst S users. It is worth noting that there are usually both statistical and computational inefficiencies to consider, and we do recommend that users consider the more sophisticated uses of the bootstrap which Davison & Hinkley (1997) expound extremely well.

Jackson (1975, p.vii) in discussing optimization in program design provides two much quoted rules

Rule 1 Don't do it.

Rule 2 (for experts only) Don't do it yet—that is not until you have a perfectly clear and unoptimized solution.

to which we might add 'to the right problem by an efficient method'. Jackson's first rule is particularly pertinent to programming the bootstrap. There is an extensive library of S functions written by Angelo Canty to accompany Davison & Hinkley (1997) (and to compute its many illustrative examples) and ports are available for all three engines. Further, recent versions of S-PLUS have a `bootstrap` function written by Charles Roosen in consultation with subject-matter experts.  Please do as we do and make use of these.  Also, as we shall see, much can be learned by studying their S code.

Simple bootstrapping computes the statistic of interest, say `statistic(x)`, applied to a subset of the same size of `x` drawn randomly with replacement, repeats this $B$ (many) times and looks at some aspect of the distribution of the bootstrapped values.  Thus a naive bootstrap function might be

```
naive.boot <- function(x, statistic, B = 200, ...)
{
  n <- length(x)
  res <- vector("list", B)
  for(i in 1:B) {
    res[[i]] <- statistic(sample(x, replace = T), ...)
    NULL
  }
  res
}
```

Note that we use a list to collect the results as we do not know what sort of object `statistic` returns, or even if that depends on the inputs.

Cross-validation is another method used to study the performance of solutions to statistical problems, and once again has many variants (Ripley, 1996, Section 2.7).  Like bootstrapping, it involves applying the original procedure to several subsets of the original data.  Except for leave-one-out cross-validation (which has many statistical disadvantages) of some statistics, the repeated applications do need to be done numerically.

A $V$–fold cross-validation function will need a function to fit the model, and a `predict` method.  For the forensic glass problem in MASS (pages 360–4) we used

```
rand <- sample (10, 214, replace = T)
CVtest <- function(fitfn, predfn, ...)
{
  res <- fgl$type
  for (i in sort(unique(rand))) {
    cat("fold ", i, "\n", sep="")
    learn <- fitfn(rand != i, ...)
    res[rand == i] <- predfn(learn, rand == i)
    NULL  # useful in S-PLUS 3.x
  }
  res
}
```

There are 214 rows in the data frame `fgl`, so `rand` randomly allocates each one with equal probabilities to one of 10 groups.  Note that this works with logical vectors of cases to include or exclude rather than pass around whole datasets.

How can we improve on these simple functions (if we need to)?  One idea is to work with indices for the bootstrap, which allows more complicated datasets than a single vector $x$, but we can always arrange this by bootstrapping the indices $1:n$ and using them inside `statistic`. Other ways to consider the bootstrap sample are as a weighted version of the original sample, either with frequencies summing to $n$ or weights summing to one, and one of these can be the most appropriate specification for some fitting algorithms. Canty's function `boot` allows all these possibilities and other ways of sampling (for example the balanced bootstrap and to specify weights for the sampling), but in essence calls a `for` loop like `naive.bootstrap`.

The S-PLUS function `bootstrap` tackles the need for other bootstrap methods by having an argument `sampler`. The default sampler is

```
samp.boot.mc <- function(n, B)
{
  x <- sample(n, B * n, replace = T)
  dim(x) <- c(n, B)
  x
}
```

which sets up a matrix of indices. We would have probably written a one-line version of this function with body

```
structure(sample(n, B * n, replace = T), dim = c(n, B))
```

The balanced bootstrap is obtained by sampling without replacement from B copies of $1:n$.

We have seen several times that there may be advantages in using one of the `apply` functions rather than `for`, and for the correlation coefficient simulation that we might want to split a large problem into smaller pieces and loop over the pieces. That is the strategy adopted by `bootstrap`: it splits the calculation to blocks, by default of size 100, uses `lapply` on each block and loops over blocks (carefully as the last block may be smaller than the others). Note that with the ordinary bootstrap this is merely a convenience for the user in producing a single summary object, as the user could call `bootstrap` repeatedly in a `for` loop at the top level (and that may well use less memory) or in a `For` loop or even run different blocks on different computers (and perhaps thereby risk the wrath of system managers).

Other aspects of `bootstrap` are worth noting to illustrate the care needed to write a robust general-purpose S function.

(a) There is an 'on exit' function to save partial results in the way we illustrated on page 47, but only after each block.

(b) Care is taken that the randomization used by the sampling does not interfere with any use of the random number stream by the user's code inside `statistic`, at least if the user requests this or if applying the function to the original data uses the random number stream. (The original data might be rather special: it probably has no ties and bootstrap resamples do.)

(c) The results are stored as a list, and this is simplified to an array only after checking every run gave a result of the same length as on the observed data. (The results are thrown away otherwise, which seems wasteful.)

(d) Optionally, the data to be used are stored in frame 1 to ensure visibility.

(e) The fitting function will be called many times, and a special-purpose function is created for the problem taking into account that the `statistic` argument could be an expression (not a function), that assignment to frame 1 might be needed, and so on.

(f) `boot` has a ... argument to pass arguments to `statistic`, but `bootstrap` does not. Allowing this is usually a good idea, and we have seen ways in Section 3.5 and earlier in this chapter to capture such arguments.

One use for bootstrapping in pattern recognition problems is known as *bagging* (Breiman, 1996) where the results from a classifier applied to a set of bootstrap resamples are averaged or aggregated. *Boosting* (Freund, 1990; Schapire, 1990; Freund, 1995; Freund & Schapire, 1995, 1996a,b) is a class of techniques that makes use of results on weighted versions of a dataset in a similar way to bagging, the difference being that the weights are chosen iteratively based on the collective performance on the runs so far. Boosting can be implemented in S quite easily: indeed BDR did so in half an hour to try some experiments with Yoav Freund and Leo Breiman.

Can we improve the cross-validation code? One thing we can think of doing is balancing the sampling, so that the 10 groups are as near equal in size as possible. This is theoretically better, but the improvement will be very modest. Interestingly, `cv.tree` uses the unbalanced approach, and the `rpart` library the balanced one. Balanced sampling is easily done as for the balanced bootstrap by

```
rand <- sample(rep(1:V, length = 214), 214, replace = F)
```

With V = 10 this gives 6 groups of size 22 and 4 of size 21. As the number of groups is usually small, a `for` loop is as good a way as any, although using a `For` loop may help avoid memory-buildup (and `cv.tree` has an option to use one).

Cross-validation is sometimes used to select tuning parameters in methods: indeed that is how it used in `cv.tree`. MASS has examples for neural networks on pages 301 and 363: indeed the latter has cross-validation for tuning inside cross-validation for testing and uses 1 000 fits taking several hours. Such replication is often what makes efficient use of CPU time important.

## 7.6 Maximum likelihood estimates and iterative calculations

Some calculations are genuinely iterative and resist vectorization. Such problems are normally best tackled using compiled code (possibly indirectly), but prototyping in S on small problems can be a great way to understand the problem and to generate reference solutions to test compiled code.

Many occur in optimization problems: for example those simple techniques beloved of statisticians, Fisher scoring and the EM 'algorithm' (Lange, 1999) move through the parameter space in a series of steps each depending on the current position. If more than a modest number of steps are required a simple S program can be too slow and/or use too much memory.

We have often found that direct maximization provides an even simpler programming task and allows us to benefit from the expertise of the writers of general-purpose optimization problems. Two examples from MASS are the fitting of a mixture of normals (pages 262–6) and logistic regression (page 269), archetypal problems for EM and Fisher scoring (or iteratively re-weighted least squares, IRLS) respectively. The advantage of a simple program are often statistical rather computational: it allows us to spend time thinking about the results and making desirable alterations to the models. Those examples provide good illustrations.

**Normal mixtures** It has been known for a long time that this is not a well-posed problem: the likelihood is unbounded and normally has many local maxima. So it pays to study the solution(s) carefully. Direct maximization and the standard EM approach are about equally easy to program. Finding the Hessian of the log-likelihood and hence approximate confidence regions for the estimates is trivial in our direct solution: it can be done in the EM approach but requires much more work. Allowing the parameters to depend on covariates was a simple exploration in the approach we took, but would need a completely separate program in the EM approach.

**Logistic regression** Fisher scoring reduces to IRLS for logistic regression, and normally works well. However, it does not always work well, as the Hauck–Donner effect (MASS p. 225) shows that the information matrix can give a rather local estimate of the shape of the log-likelihood. Our function `logitreg` (MASS p. 269) was originally written to allow non-negativity constraints to be placed on some of the regression coefficients, but has since been used as a basis for exploring several other generalizations of logistic regression.

Another class of iterative optimization problems are given the umbrella term 'dynamic programming'. Originally designed for a time series of allocation or decision problems (such as the 'secretary' problem), the principle has found much wider applicability, for example the Viterbi (1967) algorithm much used in speech recognition. An application of much current interest is to genome sequencing (Waterman, 1995, Chapter 9; Durbin *et al.*, 1998, Chapter 2). We will illustrate sequence matching, but applied to a very different field, finding similarity measures between Gregorian chants for musicology. Given two chants (or two verses of a chant), we wanted a numerical measure of their similarity. Applied to all pairs from a collection of chants this gives a similarity matrix that can be used in several ways, for example to cluster or map (by multidimensional scaling) the collection.

The musical content of a verse is a series of *neumes*, of which there are around 25 types, and the verses can be of different lengths. We seek a match that finds long matching subsequences: 'matching' is penalized for non-matches or insertions. The Smith–Waterman algorithm (Durbin *et al.*, 1998, p. 22) in S is

```
local <- function(seq1, seq2, sx=1, ss)
{
  n <- length(seq1)
  m <- length(seq2)
  S <- matrix(0, n+1, m+1)
  for (i in 1:n)
    for (j in 1:m)
    {
      if (seq1[i] == seq2[j]) sxy <- ss[seq1[i]]
      else sxy <- -2
      poss <- c(S[i, j+1] - sx, S[i, j] + sxy, S[i+1, j] - sx)
      allposs <- seq(along = poss)[poss == max(poss)]
      S[i+1, j+1] <- max(0, poss)
    }
  list(S=S, seq1 = seq1, seq2 = seq2)
}
```

This is a dynamic programming algorithm that is intrinsically recursive. It builds up the answer by finding the score for matching the first $i$ neumes of the first verse to the first $j$ of the second verse, and increases $(i, j)$ in a raster scan. The weights ss control the match scores: matching very common neumes is not rewarded as much as matching 'interesting' ones.

On an example with 6 verses and 381 neumes in total running this for all 15 pairs of verses took 15 minutes. Something as simple as this is a clear candidate for conversion to C. The C code is only slightly longer, but our version makes use of utility routines to create matrices as arrays of pointers to arrays. The C version took 4 seconds. When comparing this whole chant with another with 197 neumes, S-PLUS 2000 took an hour and 45Mb,[7] R took 4 minutes (using the standard workspace size) and the C version less than 1 second.

We *can* vectorize this calculation by re-ordering it. To calculate S[i, j] we need to know the values for (i-1, j), (i-1, j-1) and (i, j-1) so if we loop on k = i+j we can update a diagonal of S simultaneously. On the two-chant problem this took about 18 seconds and little memory in S-PLUS 2000.

## 7.7 Tips

This section is a collection of small points, each of which can help save memory usage.

1. Re-use temporary variables if appropriate. If a variable of length $n$ is used for several different purposes within a function, S will usually re-use the allocated storage. Note that this can make code much harder to read, as variables will not have meaningful names—so add some comments.

---

[7] on a machine with ample RAM; it failed to complete overnight on a machine with 64Mb of RAM.

2. Avoid growing an object. Suppose we want to keep the results in a vector res. Do *not* use

```
res <- NULL
for (iter in 1:1000)
    res <- c(res, myfunction(iter, moreargs))
```

as this may force a copy of res to be made at each iteration; rather use

```
res <- numeric(1000)
for (iter in 1:1000) {
    res[iter] <- myfunction(iter, moreargs)
    NULL
}
```

This is even more important if the result is a matrix rather than a vector. In some circumstances the final size of res may be unknown, in which case allocate a sufficient size and shrink the vector at the end of the loop.

3. Remove attributes from frequently used objects. The names attribute of a vector can take as much space as its values and the presence of dimnames can severely degrade performance in very large matrix calculations.

4. Save useful bits of computations. Good examples are logical and subscripting calculations; for example a classification tree will have

```
ctree <- !is.null(attr(tree, "ylevels"))
```

true, and this assignment can be done within the first if statement of a function which handles both regression and classification trees.

5. Conversely, do not name intermediate results unnecessarily. If an intermediate result is only needed once further down in a calculation, consider weaving it in as an expression at that point rather than as the name of an already evaluated value.

The function remove can be used to remove intermediate results that are no longer needed, but care will be needed to specify the right frame, for example (an S engine) by

```
remove("foo", frame=sys.nframe())
```

6. (S engines only.) Using get with an explicit frame or with the argument immediate=T instead of letting the parser find a data object ensures that the object is not retained in memory for possible future use. Using assign with immediate=T (see page 58) or a direct call to dbwrite saves the memory associated with backing-out an assignment.

7. Coding the innermost calculations as C or FORTRAN functions can give rise to dramatic savings in memory usage and computation time. The first edition of MASS had both C and S forms of our spatial library, but as the C version was 10–15 times faster we dropped the S version.

# Chapter 8

# S Software Development

In this chapter we consider the tools available for the process of software development in the S language. Such tools are often a matter of personal taste so many are available, and our aim is to cover all the possibilities at fairly shallow level.

The process of software development can be thought of as covering the following phases.

- *Develop* a working version of the S functions. This will involve editing functions, and almost certainly will involve *debugging* to track down errors or even to monitor that intermediate results are as expected.

- *Check* the code against examples with known answers.

- *Refine* and *polish* your code in the same way that you would polish your English prose. As you would use a dictionary, use your reference materials (the on-line help and especially the S code of system functions) to check your understanding. Other people will be reading your code and in refining it you will often find ways to improve and simplify it that you did not expect.

- *Document* your code fully, clearly and (preferably) in a way that can be verified by formal tests. This is another way of ensuring that your logic is correct and your plan is complete. Give your checks as part of the documentation so later versions can be re-checked. (R has functions to run the examples on the help pages, which is helpful if these are comprehensive enough.)

- *Package* and *distribute* your code. In S-PLUS this is normally done by distributing a library section (which R calls a package); if the code is to be used on more than one platform, the differences between platforms need to be considered. We end this chapter with a check list we have found useful.

These phases can of course occur in parallel, and documentation is often best done alongside checking.

## 8.1   Editing S functions and objects

An S object may be modified in several ways, but in all cases the object is given a text representation in an external file or window, the file is edited in some way, and the object re-created by assignment. Commonly used methods are

1. Small adjustments can often be made by cut-and-paste between the running session and an editable window. Sometimes it is helpful to save the function definition in a file (from the editor or with `dump("obj", "obj.q")`) and use `source("obj.q")` to read it in again. Note that q is the preferred filename extension in S-PLUS, and R in R.

2. The function `fix` takes as argument an object to edit:

   ```
   > fix(obj)
   ```

   This initiates an editing session with a text version of `obj` available for correction. On completion of the editing session the corrected version is assigned to the object and the session is resumed. The editor used is a system default. Under UNIX this is usually `vi` and in Windows `notepad`, but a different editor can be specified as an `options` argument:

   ```
   > options(editor = "emacs -nw")
   ```

   The most recently edited object can be re-edited by invoking `fix()` (except in R).

3. The function `ed` (`edit` under R) can achieve the same outcome by explicit assignment. It takes an object to edit as its principal argument and the name of an editor to use as another optional argument:

   ```
   > obj <- ed(obj, editor = "emacs -nw")
   ```

   If `vi` is the editor chosen we can use `obj <- vi(obj)`; under Windows `obj <- edit(obj)` uses `notepad`. Again, the function can be re-edited by invoking `ed` or `vi` or `edit` without specifying an object.

4. Under S-PLUS for Windows the function `Edit` can be used to dump the function to a script window, and running the script window will source the function. Dragging an S object from the object explorer/browser to a script window is another way to dump the function to a script window.

There are two philosophies to managing S objects during development. One, advocated by Chambers (1998, Section 6.2), is to regard the S object as the master copy, and to dump it to human-readable form only for editing. The other, which we tend to prefer, is to keep master text files of the sources, edit those and source them into S-PLUS as needed.

There are advantages and disadvantages to each strategy. The '*S objects are definitive*' approach works well when development is by one programmer on one machine and when objects can easily be dumped (unlike S-PLUS for Windows

GUI objects). It seems to us to be dangerous in R where the only easy way to save objects is by saving a whole workspace, and workspaces can become corrupted. The *'text file is master'* approach needs care to ensure that the S objects actually used are the current versions in the text files.

We find it helpful to use an editor that can be tailored to the S language and will do syntax-sensitive highlighting and formatting. Script windows in S-PLUS 4.5 and 2000 provide some help, but the ESS package for Emacs (in all its flavours) provides the most comprehensive facilities. This is available from `http://ess.stat.wisc.edu/`.

## 8.2   Tracing and debugging

Many functions are permanent objects used again and again by the programmer and other users. It is important that they be written well, free of errors and clearly documented. Removing errors is often a process requiring several phases. Although we write in terms of tracking down errors, the same process can be used to monitor the computations and to check intermediate results.

This is an area in which there are considerable differences between the various engines available; Table 8.1 shows what is available. Note though that even where a function exists in all three engines, it may behave in three different ways (and `browser` does).

### Locating and correcting errors

When an abnormal exit occurs from a function it is often useful to investigate the state of the variables when and where the error occurred. The function `traceback` will almost always be helpful, provided the S-PLUS or R process has not itself crashed.

As a simple example, consider the following error.

```
library(MASS)
lda(Sex ~ FL + CL + CW + RW + BD, crabs)
Error: Object "Sex" not found
```

The new S engine suggests 'Use `traceback()` to see the call stack', and if we do that we find

```
> traceback()
Message: Object "Sex" not found
5:
4: model.frame.default(formula = Sex ~ FL + CL + CW +
   RW + BD, data=crabs)
3: eval(m, sys.parent())
2: lda.formula(Sex ~ FL + CL + CW + RW + BD, crabs)
1:
```

**Table 8.1**: Some debugging facilities.

| Function | S | R |  |
|---|---|---|---|
| `print, cat` | ✓ | ✓ | Printing key quantities from within a function may be all that is needed to locate an error. |
| `traceback` | ✓ | ✓ | Prints the calls in process of evaluation at the time of any error that causes a dump. |
| `browser` | ✓ | ✓ | Function that may be inserted to interrupt the action and allow variables in the frame to be investigated before the error occurs. |
| `trace` | ✓ | ✓ | Place tracing information at the head of, or inside, functions. Under S, may be used to insert calls to `browser` at specified positions. |
| `untrace` | ✓ | ✓ | Turns some or all tracing off. |
| `tprint` | ✓ |  | Produces a numbered listing of the body of a function for use with the at argument of `trace`. |
| `debug, undebug` |  | ✓ | Add or remove a call to `browser` at the head of the function. |
| `options(warn=2)` | ✓ | ✓ | Changes a warning into an error, precipitating a dump. |
| `options(error=)` | ✓ | ✓ | Specifies the dump action. The default action in S is `dump.calls` which dumps the details of calls, only, but `dump.frames` can be used to cause a dump of evaluation frames. |
| `last.dump` | ✓ |  | The object in the `.Data` directory that contains a list of calls or frames after a dump. |
| `debugger` | ✓ |  | Function to investigate `last.dump` after an error. |
| `inspect` | ✓ |  | An interactive debugger. Forthcoming for S-PLUS 5.x. |

Each line in the traceback is a frame, so the hierarchy of calls to the one that gave the error should be clear. Two of the frames are unnamed: this is normally the case for frame 1. A traceback from the GUI interface of S-PLUS 4.x will have further frames associated with communicating between the GUI and the S engine.

Note that although we called the generic function `lda`, this does not appear in this listing from S-PLUS 3.4, as calls to generic functions are replaced by calls to a method using the same frame. When the new S engine is used, the hierarchy shown is

```
>  traceback()
8: eval(action, sys.parent())
7: doErrorAction("Problem: Object \"Sex\" not found", 1000)
6:
5: model.frame.default(formula = Sex ~ FL + CL + CW + RW + BD,
4: model.frame(formula = Sex ~ FL + CL + CW + RW + BD,
```

```
          data = crabs
    3: eval(m, sys.parent())
    2: lda(Sex ~ FL + CL + CW + RW + BD, crabs)
    1:
```

The trace in R is similar to that from S-PLUS 3.4, but numbered in reverse order:

```
> traceback()
[1] "eval(expr, envir, enclos)"
[2] "eval(attr(formula, \"variables\"), data,
    sys.frame(sys.parent()))"
[3] "model.frame.default(formula = Sex ~ FL + CL + CW +
    RW + BD, data = crabs)"
[4] "eval(expr, envir, enclos)"
[5] "eval(m, sys.frame(sys.parent()))"
[6] "lda.formula(Sex ~ FL + CL + CW + RW + BD, crabs)"
```

In this example the error is probably obvious from the error message. In S-PLUS we could investigate further by post-mortem debugging; see the next subsection.

To debug in more detail we could insert code to print out crucial quantities: in this example there appears to be a problem with the object m in lda.formula. We could use print or cat, but for exploratory work it is often most helpful to insert a call to browser. We insert the line browser() immediately before the call to eval in lda.formula, perhaps most easily by using fix.

```
> lda(Sex ~ FL + CL + CW + RW + BD, crabs)
Called from: lda.formula(Sex ~ FL + CL + CW . . .
b(2)>
```

The precise details of the browser vary by engine, but in all cases the prompt changes to show that it is coming from the browser and to give some indication of the frame. We can treat the prompt just like a command-line prompt, investigating objects by giving their name (which will auto-print them), performing computations and even changing objects by assignment. Typing ? in S-PLUS 3.x / 4.x will give a list of objects in the current frame:

```
b(2)> ?
1: .Auto.print
2: data
3: m
4: .Class
       ....
```

In any of the engines we can evaluate a few quantities to see what might be going wrong:

```
b(2)> names(m)
[1] ""            "formula" "data"
b(2)> m$formula
Sex ~ FL + CL + CW + RW + BD
b(2)> names(m$data)
[1] "sp"   "sex"   "index" "FL"   "RW"   "CL"   "CW"   "BD"
```

At this point we can see the problem (if we had not already), and we could correct it and continue. We quit the browser by 0 (old S engine), q or c (new S engine) or an empty line or c (R).

```
b(2)> m$formula[[2]] <- as.name("sex")
b(2)> m$formula
sex ~ FL + CL + CW + RW + BD
b(2)> 0
```

We could do the same thing in the new S engine, but we need to explicitly alter the formula method for lda rather than lda.formula. The command ? gives a list of *commands* available at the browser prompt: these include up and down to navigate the hierarchy of active frames and where to show the hierarchy, as well as stop to return to the command prompt.

In R we use ls to list the variables in the current frame:

```
> lda(Sex ~ FL + CL + CW + RW + BD, crabs)
Called from: lda.formula(Sex ~ FL + CL + CW + RW + BD, crabs)
Browse[1]> ls()
[1] "data"      "formula"   "m"        "na.action" "subset"
```

We can fix the formula as before and continue. Typing n at R's browser prompt puts it into single-step mode: thereafter a blank line causes it to execute the next command, and c to resume. (If you step into a loop, c will continue running until the end of the loop, then revert to single-step mode.)

If an object's name is masked by a browser command, as q, c and n can be, use print to examine them.

Think before inserting a call to browser in a loop that will be used more than a few times. There is no way to skip further calls to the browser, and it can be hard to stop execution of the function: in the new S engine use stop and in R use Q. Sending a user interrupt at the browser prompt does not interrupt execution of the current function. One workaround in the old S engine is to insert code like

```
userstop <- 0
browser()
if(userstop) stop("User request")
```

as then the user can change the value of userstop at the browser prompt and thereby interrupt the loop.

Various of the facilities in Table 8.1 facilitate adding debugging code. In R a call to debug effectively adds a call to browser at the head of the function named as its argument, and sets it to single-step mode. Editing the function or using undebug will remove the browser call. A two-step approach is needed to use debug on a function defined inside another function: use debug on the outer function, then from the browser prompt set debug on the inner function.

Function trace exists on all the systems. In R tracing a function reports when it is entered: untrace turns tracing off. In S-PLUS the default behaviour is to give the function call on entry, but trace can be used to report on exit as well (argument exit) or before a numbered line in the function body (argument at: tprint numbers lines[1] to make this easier). Further, what trace actually does is to insert a call to std.trace, and this can be replaced by a function or expression supplied as an argument to trace. As trace in S-PLUS alters a local copy of the function, do use untrace before editing the function.

## Post-mortem debugging

This is only available in S-PLUS, and we miss it sorely in R. When execution hits an error condition, by default enough information is recorded in the object last.dump to allow a traceback. As this is recorded in a permanent database, tracebacks from fatal errors in a batch job can be found by starting S-PLUS and calling traceback.

If the error argument to options is set to dump.frames, all the active frames are recorded in last.dump. This can result in a large[2] object, so it should be used cautiously, but it does enable complete post-mortem debugging after a batch job.

The new S engine allows the further option recover which can only be used in interactive sessions where it brings up a browser on the evaluation frame in which the error occurred. This sometimes allows simple corrections to be made and evaluation continued by the command go; another possible command is dump to dump the frames.

Function debugger is used (with no argument) to investigate last.dump. If the frames have not been dumped it gives essentially the same information as traceback. However, if the evaluation frames are available, debugger can run a browser in any of the frames and so allow detailed investigation of the variables at the time of the dump. First let us consider S-PLUS 3.x / 4.x.

```
> options(error=dump.frames)
> lda(Sex ~ FL + CL + CW + RW + BD, crabs)
Error: Object "Sex" not found
Dumped
> debugger()
Message: Object "Sex" not found

1:
2: lda.formula(Sex ~ FL + CL + CW + RW + BD, crabs)
```

---

[1] Only lines which begin sub-expressions are numbered: try tprint(tprint) to see the effect of this rule.

[2] possibly equal in size to the amount of available virtual memory.

```
3: eval(m, sys.parent())
4: model.frame.default(formula = Sex ~ FL + CL + CW + RW + BD,
   data = structure(
5:
Selection: 2
Frame of lda.formula(Sex ~ FL + CL + CW + RW + BD, crabs)
d(2)>
```

gives us a browser which we can use as before. Quitting the browser returns to
the debugger menu: selecting 0 exits from that.

In the new S engine the debugger behaves somewhat differently: it has essentially the browser interface, and we use up and down to navigate between
frames.

```
> debugger()
Message: Problem: Object "Sex" not found
browser: Frame 20
b()> where
*20: list() from 17
17: model.frame.default(formul.... from 14
14: model.frame(formula = Sex .... from 11
11: eval(m, sys.parent()) from 8
8: lda(Sex ~ FL + CL + CW + R.... from 5
5: list() from 1
2: debugger() from 1
1:  from 1
b()> up
Browsing in frame of model.frame.default(formul....
Local Variables:
.Class, .Generic, .Group, .Method, data, dots, formula,
   na.action, tj

b(model.frame.default)> up
Browsing in frame of model.frame(formula = Sex ....
Local Variables: formula

b(model.frame)> up
Browsing in frame of eval(m, sys.parent())
Local Variables: expression, local, parent

b(eval)> up
Browsing in frame of lda(Sex ~ FL + CL + CW + R....
Local Variables: .Generic, .Signature, data, dots, m, x, y

b(lda)>
b(lda)> m
model.frame(formula = Sex ~ FL + CL + CW + RW + BD,
   data = crabs)
b(lda)> names(crabs)
[1] "sp"   "sex" "index" "FL"    "RW"    "CL"    "CW"    "BD"
```

**Table 8.2**: Main options for the function `inspect`. Using `help` at the `inspect` prompt followed by any of these terms will give fuller details.

Advance evaluation:
    step – walk through expressions
    do – do expressions atomically
    complete – a loop or function
    resume – continue to next mark
    enter – descend into a function call
    quit – abandon evaluation

Display:
    where – current calls and expr'n
    objects – local frame objects
    show – installed tracks and marks
    find – location of an S-PLUS object
    return.value – function return value
    on.exit – scheduled on.exit expr'ns
    fundef – function definition

Halt evaluation; track functions:
    mark, unmark – arrange to stop in
      a function or at an expression
    track, untrack – install or change
      function call reporting

Examine other current frames:
    up, down

Miscellaneous:
    eval – S-PLUS expressions you type in
    help – syntax and description
    debug.options – option settings

Informational help entries:
    names – when to quote
    keywords – list of reserved words

```
b(lda)> q
NULL
```

Sometimes it can be very difficult to track down where warnings are occurring. Setting `options(warn=2)` turns them into errors, whose location can be found by `traceback`, and investigated further by `debugger`.

## An interactive debugger

S-PLUS 3.x, 4.x have (and some versions of 5.x will have) an interactive symbolic debugger, `inspect`, similar in spirit to those used for traditional programming languages. Its main options are shown in Table 8.2, which is similar to the screen obtained by a command `help` inside `inspect`.

To run `inspect`, give it an expression to inspect. In the old S engine we can use

```
> inspect(lda(Sex ~ FL + CL + CW + RW + BD, crabs))
entering function lda
stopped in lda (frame 3), at:
        if(is.null(class(x)))
                class(x) <- data.class(x)
d> mark lda.formula
entry mark  set for lda.formula
exit mark(s) set for lda.formula
d> resume
entering function lda.formula
stopped in lda.formula (frame 3), at:
        m <- match.call(expand.dots = F)
```

```
d>
d> do  (several times)
stopped in lda.formula (frame 3), at:
        m <- eval(m, sys.parent())
d> eval m$formula
Sex ~ FL + CL + CW + RW + BD
d> resume
Error: Object "Sex" not found

Calls at time of error:

7: error = function() from 6
6:   from 1
5: model.frame.default(formula = Sex ~ FL + CL + CW + RW + BD,
   data = structure( from 1
4: eval(m, sys.parent()) from 3
3: lda.formula(Sex ~ FL + CL + CW + RW + BD, crabs) from 1
2: inspect(lda(Sex ~ FL + CL + CW + RW + BD, crabs)) from 1
1:   from 1
   . . . .
```

We get a browser-like prompt: we need to remember this is not a browser, and expressions must be preceded by eval. We can step through examining variables, but when we reach the error we get a traceback and are dropped out of the debugger (and this gives an error when attempting to dump the frames).

## 8.3   Creating on-line help

Functions or data frames that may be used by more than one user, or even by one user over a long period of time, should be documented. This is easy to do in a machine readable form, but the forms for the various engines and platforms are quite different. In all cases using the prompt function with the object to be documented as argument will generate an outline help document file, which should be completed in a text editor. What the format for that file is differs radically by platform.

S-PLUS 3.x, 5.0 on UNIX: The help file is written in a nroff / troff macro language. The files are stored in the .Help subdirectory of the .Data directory, and processed at run time by nroff to produce on-screen help, and by troff to produce printed help. The source help files created by prompt have extension .d, but are stored in .Help with no extension. Some platforms (SGI) have pre-processed text help files in the .Cat.Help subdirectory of .Data.

S-PLUS 4.x on Windows: There are two possibilities:

(i) The system help files are Windows help files, and these can also be created for user-contributed libraries.

    (ii) Help can be supplied as text files stored in the _Data\_Help direc-
        tory. This is complicated by the fact that the help files have to be
        stored under the *mapped* file name of the S object.

S-PLUS 5.1 on UNIX: The help files are written in a specified dialect of SGML,
    and translated at run time to HTML for on-screen help.

R: The help file is written in a TEX–like macro language, and processed by Perl
    scripts. In R there are only facilities for adding documentation to packages,
    but then workspaces are intended for transient objects. R help files have
    extension .Rd and are stored in the man directory of a library distribution.
    They are converted to text, HTML, LATEX and source files of the examples
    when a package is installed.

This is a very wide range of possibilities: Figures 8.1, 8.2 and 8.3 show example help files for ucv. Using prompt gives an outline, and it is easy to provide
the required information using only the instructions in the prompt file itself. Further information can be obtained from the prompt help document on the system
in use. The ESS package for Emacs (see page 181) has modes to edit .d and
.Rd files.

## Keywords

A S-PLUS nroff help file ends with one or more .KW lines, then .WR. The .KW
lines denote keywords that are used to organize the information into categories in
the help.start help system. The list of keywords and their category names are
given in the file

    $SHOME/splus/lib/X11/keywords

Some of the more useful keywords are (sorted by the category)

```
aplot          Add to Existing Plot
cluster        Clustering
dplot          Computations Related to Plotting
sysdata        Data Sets
hplot          High-Level Plots
algebra        Linear Algebra
math           Mathematical Operations
multivariate   Multivariate Techniques
nonlinear      Non-linear Regression
nonparametric  Nonparametric Statistics
optimize       Optimization
print          Printing
distribution   Probability Distributions and Random Numbers
regression     Regression
tree           Regression and Classification Trees
robust         Robust/Resistant Techniques
smooth         Smoothing Operations
htest          Statistical Inference
```

```
.BG
.\" function name
.FN ucv
.\" help title
.TL
Unbiased Cross-Validation for Bandwidth Selection
.\" description
.DN
Uses unbiased cross-validation to select the bandwidth of a
Gaussian kernel density estimator.
.\" usage
.CS
ucv(x, nb=1000, lower, upper)
.\" required arguments
.RA
.AG x
a numeric vector
.\" optional arguments
.OA
.AG nb
number of bins to use.
.AG "lower, upper"
Range over which to minimize.   The default is almost
    always satisfactory.
.\" return value
.RT
a bandwidth.
.SH REFERENCE
Scott, D. W. (1992)
.ul
Multivariate Density Estimation: Theory, Practice,
.ul
and Visualization.
Wiley.
.\" see also
.SA
'bcv', 'width.SJ', 'density'
.\" examples
.EX
ucv(faithful$eruptions)
.\" keyword
.KW dplot
.WR
```

**Figure 8.1**: Help file ucv.d for ucv in S-PLUS 3.x on UNIX. The macro .\" introduces a comment, added here to make the file more human-readable.

```
<!doctype s-function-doc system "s-function-doc.dtd" [
<!entity % S-OLD "INCLUDE">] >
<s-function-doc>
<s-topics>
   <s-topic>ucv</s-topic>
</s-topics>
<s-title>
Unbiased Cross-Validation for Bandwidth Selection
</s-title>
<s-description>
Uses unbiased cross-validation to select the bandwidth of a Gaussian
kernel density estimator.
</s-description>
<s-usage> <s-old-style-usage>
ucv(x, nb=1000, lower, upper)
</s-old-style-usage> </s-usage>
<s-args-required>
<s-arg name="x">
a numeric vector
</s-arg>
</s-args-required>
<s-args-optional>
<s-arg name="nb">
number of bins to use.
</s-arg>
<s-arg name="lower, upper">
Range over which to minimize.  The default is almost always satisfactory.
</s-arg>
</s-args-optional>
<s-value>
a bandwidth.
</s-value>
<s-section name=" REFERENCE">
Scott, D. W. (1992)
<it>Multivariate Density Estimation: Theory, Practice, and
Visualization. </it> Wiley.
</s-section>
<s-see>
<s-function name="bcv.sgm">bcv</s-function>,
<s-function name="width.SJ.sgm">width.SJ</s-function>,
<s-function name="density.sgm">density</s-function>
</s-see>
<s-examples> <s-example type = text>
ucv(faithful$eruptions)
</s-example> </s-examples>
<s-keywords>
<s-keyword>dplot</s-keyword>
</s-keywords>
<s-docclass>
function
</s-docclass>
</s-function-doc>
```

**Figure 8.2**: Help file ucv.sqm for ucv in S-PLUS 5.1, slightly reformatted.

```
\name{ucv}
\alias{ucv}
\title{Unbiased Cross-Validation for Bandwidth Selection}
\description{
  Uses unbiased cross-validation to select the bandwidth of
  a Gaussian kernel density estimator.
}
\usage{ucv(x, nb=1000, lower, upper)}
\arguments{
  \item{x}{a numeric vector}
  \item{nb}{number of bins to use.}
  \item{lower, upper}{Range over which to minimize. The default
    is almost always satisfactory.}
}
\value{a bandwidth.}
\references{
Scott, D. W. (1992)
\emph{Multivariate Density Estimation: Theory, Practice,
  and Visualization.} Wiley.}
\seealso{
\code{\link{bcv}}, \code{\link{width.SJ}}, \code{\link{density}}
}
\examples{
data(faithful)
ucv(faithful$eruptions)
}
\keyword{dplot}
```

**Figure 8.3**: Help file ucv.Rd for ucv in R.

```
models          Statistical Models
ts              Time Series
utilities       Utilities
```

R has a similar system using the \keyword macro, with a standard list of keywords in the file

'R RHOME'/doc/KEYWORDS

All of those listed here are also in R but datasets is used for statistical data sets rather than sysdata.

## Using help on S-PLUS 3.x UNIX machines without nroff

Some UNIX platforms, principally SGI, are shipped without the nroff program which is used by S-PLUS 3.x to convert the help files into a readable form for the on-line help system. The system help files can be read, as pre-processed versions are stored in the directory .Cat.Help parallel to the .Help directory. This workaround can be used for users' help files too, provided that the corresponding entry in the .Help both exists and is older than that in .Cat.help.

The simplest way to create pre-processed help files is to run

```
Splus CATHELP .Data
```

which processes the files in `.Data/.Help` to `.Data/.Cat.Help`.

If you have `nroff` or a clone such as `groff` and the target users might not, consider distributing the `.Cat.Help` directory.

## S-PLUS 4.x on Windows

It is possible to create simple ASCII help documents that are often all that is needed for many functions, at least as a temporary measure. A call to the prompt function such as

```
> prompt(myfun)
```

will create an outline help file for the object in the directory `_Data\_Help` using the same *mapped* file as that used by the object itself in the `_Data` directory. The mapped file name is reported in the command window. The user should then complete the template file using a text editor such as `notepad`.

If a correction is needed to the help file but the mapped file name has been forgotten it may be found by using

```
> true.file.name(myfun)
```

### *Creating standard* Windows *help files*

The Windows help files used by the system and in our libraries are standard Windows help files which were converted automatically from the UNIX help files. A set of tools to do this is available in the file `spwinhlp.zip` at

```
http://www.stats.ox.ac.uk/pub/SWin/
http://lib.stat.cmu.edu/DOS/S/SWin/
```

This `zip` archive contains a Perl script that can be used on a UNIX or Windows system to convert a set of `nroff–style` help files to Microsoft's *rich text format*. This can be edited in a Windows word processor if desired. The `.rtf` help file is then compiled to a `.hlp` file using Microsoft's help compiler.[3]

Of course, Windows users will not usually be starting with `nroff–style` files. Our tools provide two routes to make such files. One is a Perl script which converts the ASCII files generated by `prompt` to `nroff–style`. Such files will not have the usual enhancements (such as using italic and bold in references) but these can be added by editing the `nroff–style` files or by editing the `.rtf` file in a Windows word processor. Do ensure that names of S objects are quoted (between left and right single quotes) before conversion.

The second route is a function `uprompt` that works in the same way as `prompt` under UNIX, providing a skeleton `nroff` file for completion. This will be the preferred route for experienced writers of help files on UNIX (and probably for no one else).

---

[3]This is supplied with most Windows compilers, and a version is also available on the Internet. Details are given in `spwinhlp.zip`.

## 8.4   S-PLUS libraries

In S-PLUS the official terminology is that MASS is a library *section* contained in
a directory, the *library*. R calls library sections *packages*, but they are used in
exactly the same way.

### Creating a S-PLUS library section

If the library uses no compiled C or FORTRAN code the process is fairly simple.

1. Create a directory in the library with the section name, and create .Data
   and .Data/.Help subdirectories (_Data and _Data\_Help in Windows).
   In the new S engine use Splus5 CHAPTER -m in the directory.
2. Using this as the working directory, run S-PLUS and create the desired
   objects.
3. Create help files using prompt (see page 188).
4. Create a README file (README.TXT in Windows) describing briefly the
   purpose of the library and with one-line descriptions of the public objects.
5. If any start-up actions are needed, put these in a .First.lib function.

Otherwise, the following subsections give more detailed procedures.

### UNIX, S-PLUS 3.x

1. Create a directory in the library with the section name.
2. Change to that directory and create .Data and .Data/.Help subdirectories.
3. Run S-PLUS and create the desired objects, which will be saved within the
   .Data directory. Often this is best done by

   ```
   $ Splus < section.q
   ```

   where section.q is a file listing the functions and containing expressions
   to generate the datasets. (The suffix .q is recommended to avoid any confusion with assembler code, which often has suffix .s on UNIX systems.)
4. Use the prompt function within S-PLUS to create templates of help files
   for the objects in the library which will be publicly accessible.
5. Edit the *.d files to create the help files. Copy the *.d files to the
   .Data/.Help directory *without* the .d extension, by

   ```
   $ Splus HINSTALL .Data/.Help *.d
   ```

   or by hand. For example, using sh,

```
for f in *.d
do
  cp $f .Data/.Help/`basename $f .d`
done
```

If a help file refers to several objects (by multiple .FN lines) link files with the names of the subsequent objects to the first one. (This is done automatically by HINSTALL.)

If needed, add pre-processed help files in .Data/.Cat.Help (see page 192) by

```
$ Splus CATHELP .Data
```

6.  Create a README file in the main directory describing briefly the purpose of the library and with one-line descriptions of the public objects.

7.  Add a one-line description of the library to the README file in its parent library directory, if you have write permission.

8.  Compile and link as needed any dynamically-loadable modules within the main directory (see Appendix A).

9.  Create a .First.lib object to dynamically load any modules as required. Recent S-PLUS versions (version 3.3 on some platforms and all versions of 3.4) have a function dyn.load.lib which can be used to simplify the S code to load modules. The .First.lib function can be as simple as

    ```
    .First.lib <- function(...)
        dyn.load.lib(..., basename="myname")
    ```

    This will choose dyn.load, dyn.load.shared or dyn.load2 as appropriate for the run-time platform. However, the appropriate object module must have been made available, this being myname.so for SGI and DEC Alpha, and myname_l.o for all other UNIX platforms.

10. Use help.findsum(".Data") in S-PLUS to create an index for use by help.start, or use

    ```
    $ Splus help.findsum .Data
    ```

    at the UNIX prompt.

11. Tidy up by rm -f .Data/.Audit .Data/.Last.value.

*Distributing the library*

As the binary files (both the files in the .Data directory and any compiled modules) differ between platforms under UNIX, the convention is to distribute the source files together with instructions as to how to recreate the library section.

There are two approaches. One, exemplified by our software, is to write a shell script or Makefile to perform most of the steps. We turned to shell scripts when we found difficulties in writing Makefiles that worked on all the S-PLUS platforms. The scripts need to cope with producing compiled objects for both `dyn.load` and `dyn.load.shared` and a suitable `.First.lib` function. See the on-line sources of our libraries for examples.

The other approach is to use

```
Splus CHAPTER arguments
```

to create a `Makefile` on the target platform. Consider the steps needed for our nnet library, which is distributed in source form. Running

```
Splus CHAPTER *.q *.c *.d
```

creates a `Makefile`. Then we can use

```
$ make install
     ....
$ make dyn.load
     ....
$ make clean
     ....
```

Replace `make dyn.load` by `make dyn.load.shared` if appropriate (or even `make static.load`). This will produce an object in the form expected by `dyn.load.lib`, but it will *not* write a `.First.lib` function (and many users forget to do so). The standard `Makefile` does use the name of the current directory as the section name, so on recent systems

```
.First.lib <- function(...) dyn.load.lib(..., basename="mysect")
```

will suffice.

Even after the 'clean', several unneeded files are left.


## UNIX, S-PLUS 5.x

The process here is a little simpler than under earlier UNIX versions. We dump and re-boot the chapter to store all the S objects in a few indexed files.

1. Create and move to a directory in the library with the section name.

2. Create a chapter by `Splus5 CHAPTER filename filename ...`, where the files specified are those to be compiled (and wildcards can be used).

3. Source the files containing S code, for example by

```
$ cat *.q | Splus5
```

4. Dump the chapter by `Splus5 make dump` and remove the .Data directory.

5. Run

```
$ Splus5 CHAPTER filename filename ...
$ (BOOTING_S="TRUE"; export BOOTING_S; Splus5)
$ rm all.S *.Sdata DUMP_FILES
$ Splus5 make
```

6. How the help files should be installed differs between versions 5.0 and later. A unified procedure might be

   (a) Install the help files by (using sh)

   ```
   mkdir .Data/.Help
   for f in *.d
   do
        cp $f .Data/.Help/`basename $f .d`
   done
   ```

   (b) (5.0) If a help file refers to several objects (by multiple .FN lines) link files with the names of the subsequent objects to the first one.

   (c) (5.1) Convert the help files to the new-style SGML format by either starting S-PLUS and running

   ```
   convertOldDoc(".", ".")
   ```

   or running the following as part of a script

   ```
   for f in .Data/.Help/*
   do
      $SHOME/cmd/doc_to_S $f > .Data/__Shelp/`basename $f`.sgm
   done
   ```

   Finally remove the old-style help by rm -rf .Data/.Help. Then either change the permissions on .Data/__Hhelp to allow all users to write to it, or pre-process all the SGML files to HTML by

   ```
   Splus5 CATHELP_SGML .Data
   ```

   This is also a useful check on the conversion process.

7. Tidy up by rm -f .Data/.Audit .Data/.Last.value.

*Distributing the library*

We distribute our libraries as source code and use a shell script to install it. See the on-line sources of our library for current examples.

## Windows, S-PLUS 4.x and 2000

1. Create a directory in the library with the section name.

2. Change to that directory and create _Data and _Data\_Help subdirectories.

3. Run S-PLUS and create the desired objects, which will be saved within the .Data directory. Often this is best done by a BATCH command using a file section.q listing the functions and containing expressions to generate the datasets. (The suffix .q is conventional for S code.)

4. Use the prompt function within S-PLUS to create templates of help files for the objects in the library which will be publicly accessible.

5. The templates are put directly into the _Data\_Help directory, and may have mapped names of the form __27. Edit these files in your favourite editor.

   (Optional.) Convert these help files to a Windows help file (see page 193).

6. Create a file README.TXT in the main directory describing briefly the purpose of the library and with one-line descriptions of the public objects.

7. Add a one-line description of the library to the file README.TXT file in its parent library directory if you have write permission on that file.

8. Compile and link as needed any dynamically-loadable modules and DLLs within the main directory.

9. Create a .First.lib object to dynamically load any modules as required. The .First.lib function can be as simple as

```
.First.lib <- function(...)
    dyn.load.lib(..., basename="myname")
```

However, the appropriate object module must have been made available, this being myname.obj under Windows, and that needs to be created by the obsolete Watcom compiler. For a library that depends on a DLL, the .First.lib function will need to be something like

```
.First.lib <- function(section,library) {
    path <- paste(library, section, "name.dll", sep="/")
    syms <- c("symbol1,", "symbol2") # etc
    dll.load(path, call="cdecl", symbols=syms)
}
```

10. Launch S-PLUS, attach the library and use make.DB.summary(n) (where n is the position at which the library is attached) to update the summary files _Data\__sum4.txt and _Data\__sum4i.txt.

*Distributing the library*

Library sections for Windows can be distributed in binary form. The convention is to use a .zip file. We do this from the library directory (the parent of the section) by using

```
zip -r section.zip section
```

Large numbers of small files in the _Data directory can waste enormous amounts of space, and it is more efficient to use __BIG indexed files for S-PLUS objects, as the system itself does. There is an undocumented .Internal function which can be used to generate these files. We can make use of this by

```
wbigfile <-
  function(where = 1, o = objects(), directory = ".",
           bigfile = paste(directory, "__BIG", sep = "/"),
           bigfile.index = paste(directory, "__BIGIN", sep = "/"))
{
  data <- vector(mode = "list", length = length(o))
  for(i in seq(along = o))
    data[[i]] <- get(o[[i]], where = where, immediate = T)
  names(data) <- o
  .Internal(put.to.bigfile(data, bigfile, bigfile.index,
    "standard"), "S_put_to_big_file", T, 0)
}
```

which takes a list of objects and produces a single file __BIG with index file __BIGIN. This can be used to convert a library by the following batch file (run from the main directory of the library section), called tobig.bat in our on-line scripts.

```
rem > null.q
rem set SHOME, S_WORK and path as needed
sqpe < convert.q
move _Data _Data.old
mkdir _Data
move __BIG _Data
move __BIGIN _Data
rem > _Data\__db3.1
move _Data.old\_Help _Data
start /w /min splus.exe BATCH null.q
del null.q
```

where convert.q contains

```
library(wbigfile)
if(exists("last.dump")) rm(last.dump)
if(exists(".Last.value")) rm(.Last.value)
wbigfile()
```

and wbigfile is in library wbigfile. Note that the _Data directory needs to contain an empty file named __db3.1.

Step 10 needs to run after the conversion to build the summary files used by the object browser/explorer.

## 8.5   R packages

A source package in R contains files DESCRIPTION, TITLE, INDEX and may
have directories R, data, src and man (and possibly others). The R source
files go in R and have one of the extensions .R, .S, .q, .r or .s. C
and FORTRAN source files go in src, together with a Makefile if required.
The man directory should contain R documentation files with a .Rd extension.
Datasets are stored in data directory as R code (.R or .q), matrices to be read
by read.table(file, header=T) (.tab, .txt or .csv) or saved R code
(.rda).

When a package is attached, the function .First.lib is called if this exists
in the package. This is typically used to load compiled code, by something like

```
.First.lib <- function (lib, pkg)
   library.dynam("MASS", pkg, lib)
```

Note the different orders of the arguments!

Once the main files have been put in the appropriate directories, the top-level
files can be created. Use

```
R CMD Rdindex man/*.Rd > INDEX
```

to create the INDEX file, create a suitable one-line file TITLE for use by
library() and a DESCRIPTION file along the lines of

```
Package: mytest
Version: 0.2-1
Author: An author <author@dept.domain.dom>.
Description: Something about the contents
License: GPL (version 2 or later)
```

Include a Depends: line if this package depends on other packages or on a particular version of R.

Anything extra that you want to be installed (for example, copyright information or extra documentation) should be put in a inst directory.

For further details see the R manuals. The process of creating the sources for
a package is the same under all platforms, but distribution and installation differs
on Windows.

### Distributing the package – UNIX

All that is needed is to package up the source files, usually as a gzip-ed tar
archive. The package can then be installed by

```
R INSTALL pkg_version.tar.gz
```

or by unpacking it and using R INSTALL with the directory name.

## Distributing the package – Windows

Under Windows it is conventional to distribute compiled packages that can be unpacked directly into the library directory. (To build a package from source you will need the R source-package distribution, the appropriate compilers and Perl, plus further tools to build the help files.)

For each package you want to install, unpack it to a directory, say mypkg, in R_HOME\src\library, then

```
cd R_HOME\src\gnuwin32
make pkg-mypkg
cd R_HOME\library
zip -r9X mypkg.zip mypkg
```

and distribute mypkg.zip.

## Checking the package

It is important to check a package before distributing it, and R provides some help in doing so. The commands

```
R cmd check [-l lib.loc] mypkg
```

(UNIX) and

```
make pkgcheck-mypkg
```

(Windows) will run all the examples from all the help files, and report if they ran successfully.

## 8.6  Developing code to be used on more than one engine

Writers of libraries are often faced with maintaining code that can be used on two or more of the engines. We have done so with our libraries for some time and now try as far as possible to use a single code base. For S code we use a cpp–like preprocessor written in Perl (and supplied with the scripts for this chapter) that enables us to use constructs such as (from negbin.q in library MASS)

```
#ifdef SP5
invisible(setOldClass(c("negbin", "glm", "lm")))
invisible(setOldClass(c("summary.negbin", "summary.glm")))
#endif
    ....
#ifdef R
    s <- order(-dflis)
#else
    s <- sort.list(-dflis)
#endif
```

and which automatically converts oldClass to class and so on.

For C code we can use the macros such as USING_R defined on page 128.

## 8.7   A checklist

Here is a list of common mistakes that are so easy to make that we check our own code against it before releasing it. We leave it to the reader to find a favourite way to do this: we use `grep` on the source code or a regular-expression search in an editor.

- Did you really mean `max` and `min` with more than one argument rather than `pmax` and `pmin`? The function `max` is equivalent to concatenating (applying `c` to) the arguments and then finding the largest value in a single vector. It may be better to use explicitly `max(c(arg1, arg2))` for future reference.

- Any comparison with `NA`, for example `x == NA`, should be written using `is.na`. An especially confusing version is `x == "NA"`, which coerces `x` to character, and so is false for `NaN` (see page 22).

- Check that `&` should not be `&&` and *vice versa*, and that `|` and `||` are not confused.

- The construction `1:length(x$y)` will give `c(1,0)` if `x$y` is empty or `x` does not have component `y`. Almost always `seq(along=x$y)` is better.

- Watch the precedence of `:` which comes below `^` and unary minus but above the other arithmetic operators. Use parentheses copiously.

- Consider if `!is.null(x)` or `length(x) > 0` is the better test. The difference is that empty vectors are not null. The most common idiom is to test for the existence of a component of list: `is.null` is appropriate as `x$y` is null if the component does not exist.[4]

- Check the behaviour you want when subscripting matrices and arrays. The default is to drop dimensions of size one, which can result in a vector. For safety, add `drop=F` or `drop=T` to all such calls.

- Check that when `dim` is called its argument is known to be a matrix, array or data frame.

- Ensure that the result of the last statement in a `for` loop is `NULL`. In S-PLUS 3.x this avoids the last expression of the loop being retained in memory. In other engines it is not needed, but is harmless.

- Check that method functions have exactly the same set of arguments as the generic function *including* any ... argument.[5]

- Avoid formal argument names such as `print` and `plot` and other common function names. Use a name like 'plot.it' or even 'plot.' which partial matching will allow the user to abbreviate to 'plot'. This avoids warnings in S-PLUS 3.x and also works around some bugs.

---

[4]It is also true if the component exists but is literally `NULL`, but such components are rare: for example `x$y <- NULL` removes component `y` rather than setting it to `NULL`. If this difference matters, you could use `!is.na(pmatch("y", names(x)))`.

[5]to mop up arguments given to future methods that might call this one.

- Do not rely on partial matching of arguments or names of list components, but use the full name. This avoids any problems with future extra arguments or lists with more components than you expected.

- Check that the storage modes of *all* arguments passed to .C or .Fortran are known, probably by coercion (using as.double) or perhaps by setting storage.mode. Here is an example of what can go wrong, from S-PLUS 3.4.

```
> library(MASS)
> for(df in 2) print(gam(log(perm) ~ area + peri +
    s(shape, df=df), data=rock))

Degrees of Freedom: 48 total; 42.99963 Residual
Residual Deviance: 30.61766
> for(df in 2:3) print(gam(log(perm) ~ area + peri +
    s(shape, df=df), data=rock))

Degrees of Freedom: 48 total; -1072707733 Residual
Residual Deviance: 31.92164
    . . . .
```

The problem is that 2:3 gives df storage mode "integer", and the author of s.wam did not check the mode in the call to the FORTRAN function backfit.

This is much more pressing in the new S engine, where it is much easier to get vectors of storage mode "integer".

- Remember the scope rules (pages 60 and 64); in particular consider if your functions will work when called from within a function, then test that they do.

Within a function, a call to eval probably needs a non-default value for local (S) or envir (R).

Functions defined within functions do not have automatic access to local variables of the parent in the S engines.

Functions such as tapply, apply and nlminb make provision for extra arguments to be passed down to their function arguments: they are needed more often than is commonly realised.

- Use masked or conflicts to check that any conflicts of names are deliberate.

- If you have compiled code and it might be used on an S system with dyn.load or dyn.load2, check the symbol names (see page 149). In any case, check your code is not exporting symbols with names that are likely to conflict with other users' code.

# Chapter 9

# Interfaces under Windows

S-PLUS 4.x for Windows introduced a new way to program menus and dialogs. In essence the system programs the Axum engine on which the 4.x GUI is based. Even those who much prefer the command-line interface (including the authors) may find they need to develop a graphical user interface to their functions to enable other people to make use of them.

Programming for the GUI was not initially particularly well documented, and most of the techniques shown here were discovered by browsing through the system examples and by trial-and-error. We describe the system as implemented in S-PLUS 2000; most of the examples work in 4.5 and rather fewer in 4.0.

The major step in setting up the user interface is to design a dialog box from which to launch the function. This can then be linked into the menu system and made available as a context-sensitive menu in the object explorer/browser for objects of certain classes.

It will be helpful to understand where the GUI information is stored. The interface objects are usually referred to as *objects* or *properties* and are stored in the user's preferences files, in the _Prefs directory of the project, as are customized menus. Thus a customized GUI needs to be associated with a project rather than a library. We discuss ways to manage the GUI objects in Section 9.3.

Another way to build a user interface is to interact with S-PLUS from another program, using DDE or ActiveX (formerly OLE) Automation. This can be a very powerful approach, and we discuss some techniques in Sections 9.4 and 9.5. These too are the result of extensive experimentation. See Figure 9.7 on page 231 for a taste of what can be done.

R comes as source code and with details of how to interface to its DLL under Windows. Section 9.6 gives an overview of some interfacing possibilities.

## 9.1 Building a dialog box

The main step is to associate a dialog box with a special-purpose S function. Then when this function is invoked from a menu or a toolbar button or (by double-clicking) in an object explorer/browser the dialog box is launched; when completed it supplies the arguments to the S function. This in turn will call standard functions and apply the requested actions to the results. Conventionally such functions have names that begin with `menu` such as `menuLm` and the system examples are in library `menu`. The full source for the examples here is in the file `dialogs.q` in our library `MASSdia`, part of the on-line library bundle for MASS for S-PLUS 4.x. The MASS library for S-PLUS 2000 has similar functions; please use those rather than type in the examples here.

## Defining the dialog box

Let us consider how to set up a special-purpose function interface to `lda`. The function might be defined as

```
menuLDA <-
  function(formula, data, subset, na.omit.p = T,
    method = "moment", newdata = NULL, predict.save = NULL,
    predict.method = "plug-in", plot.p = F, plot.dimen = 99,
    CV = F, nu = 5)
{
  fun.call <- match.call()
  fun.call[[1]] <- as.name("lda.formula")
  if(na.omit.p) fun.call$na.action <- as.name("na.omit")
  else fun.call$na.action <- as.name("na.fail")
  fun.args <- is.element(arg.names(fun.call),
      c("formula", "data", "subset", "na.action",
        "method", "CV", "nu"))
  fun.call <- fun.call[c(T, fun.args)]
  ldaobj <- eval(fun.call)
  if(!CV) tabResults.lda(ldaobj)
  if(!CV)
    tabPredict.LDA(ldaobj,newdata,predict.save,predict.method)
  if(!CV && plot.p) tabPlot.LDA(ldaobj, plot.dimen=plot.dimen)
  invisible(ldaobj)
}

tabResults.LDA <- function(object)
{
  cat("\n\t*** Linear Discriminant Analysis ***\n")
  invisible(print.lda(object))
}

tabPredict.LDA <-
  function(object, newdata = NULL, predict.save = NULL,
           predict.method = "plug-in")
{
  if(!is.null(predict.save)) {
    if(is.null(newdata))
      predobj <- predict(object, method = predict.method)
    else
      predobj <- predict(object, newdata = newdata,
```

```
                              method = predict.method)
      if(exists(predict.save, where = 1)) {
        newsave.name <- unique.name(predict.save, where = 1)
        assign(newsave.name, predobj, where = 1)
        warning(paste("Predictions saved in", newsave.name))
      } else assign(predict.save, predobj, where = 1)
    }
    invisible(NULL)
}

tabPlot.LDA <- function(object, plot.dimen = 99)
{
    plot.lda(object, dimen = plot.dimen)
    invisible(NULL)
}
```

The function is called when the OK or Apply[1] button on the dialog box is selected. By default a dialog box will be generated that allows the arguments to be
filled in. Figure 9.1 shows the default dialog box for our menuLDA. This dialog
box can be invoked in several ways, for example by highlighting menuLDA in
a script window and selecting Show Dialog... from the right-click menu, or by
double-clicking on menuLDA in the object explorer or object browser or by the S
command

```
guiDisplayDialog("Function", "menuLDA")
```

Filling it in and clicking on the OK or Apply button will invoke the function and
put the result in the object named in the field Return Value. If Apply is clicked
the dialog box is left up, but if the user clicks on OK or hits the return key the box
will disappear.

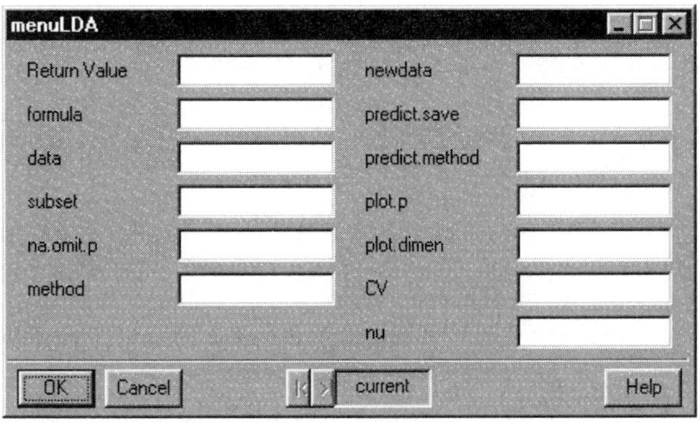

**Figure 9.1**: Default dialog box for menuLDA.

Often the main point of writing a GUI interface is user-friendliness, and Figure 9.1 is not particularly welcoming so we will write our own.

---
[1]Only some dialog boxes have Apply buttons; those to be used only once (including automatically
generated boxes) do not.

*Specifying the dialog box to be used*

A customized dialog box is selected by specifying a `FunctionInfo` GUI object. For our example we might use

```
guiCreate("FunctionInfo", Name = "menuLDA",
         Function = "menuLDA",
         HelpCommand = "help(lda)",
         DialogHeader = "Linear Discriminant Analysis",
         PropertyList = c("ldaModelPage", "ldaPlotPage",
           "ldaPredictPage", "SPropPFEnableButton"),
         CallBackFunction = "backLDA",
         ArgumentList = "#0=ldaSaveAs, #1=SPropPFFormula,
   #2=SPropDataFrameList, #3=SPropSubset, #4=SPropOmitMissing,
   #5=ldaFitMethod, #6=SPropPredictNewdata,
   #7=SPropSavePredictObject, #8=ldaPredictMethod,
   #9=ldaPlot, #10=ldaPlotDimen, #11=ldaCV, #12=ldaFitNu")
```

The result will be the multi-tabbed box shown in Figure 9.2, once we have defined all the GUI objects mentioned.

Function `guiCreate` is a call to the GUI programming language, and refers to other GUI objects, which are either created by `guiCreate` or (as in `SPropPFEnableButton`) are already defined. (All of these further GUI objects are also stored in the project's _Prefs directory, even the system ones. All the objects referred to are *properties*.) The `HelpCommand` argument gives the S function which will be invoked when the Help button of the dialog box is selected.

A dialog box is built up hierarchically. At the top level our `FunctionInfo` object defines the dialog as being made up of three tabbed pages and the standard bottom row of buttons (which cannot be omitted and are not explicitly specified). Each page is then normally made up of *groups* (or wide groups), surrounded by box on the dialog box page. For example, the `ldaModelPage` tab is defined by

```
guiCreate("Property", Name="ldaModelPage", Type="Page",
         DialogPrompt="Model",
         PropertyList="ldaModelData, SPropFSpace1, SPropFSpace2,
           SPropPFFormulaG, ldaMethod, ldaSaveModel")
guiCreate("Property", Name="ldaModelData", Type="Group",
         DialogPrompt="Data",
         PropertyList="SPropDataFrameList, SPropSubset,
           SPropOmitMissing")
guiCreate("Property", Name="ldaMethod", Type="Group",
         DialogPrompt="Fit Method",
         PropertyList="ldaFitMethod, ldaCV, ldaFitNu")
guiCreate("Property", Name="ldaSaveModel", Type="Group",
         DialogPrompt="Save Model Object",
         PropertyList="ldaSaveAs")
```

The groups contain the basic property items, so that for example the `ldaMethod` group is

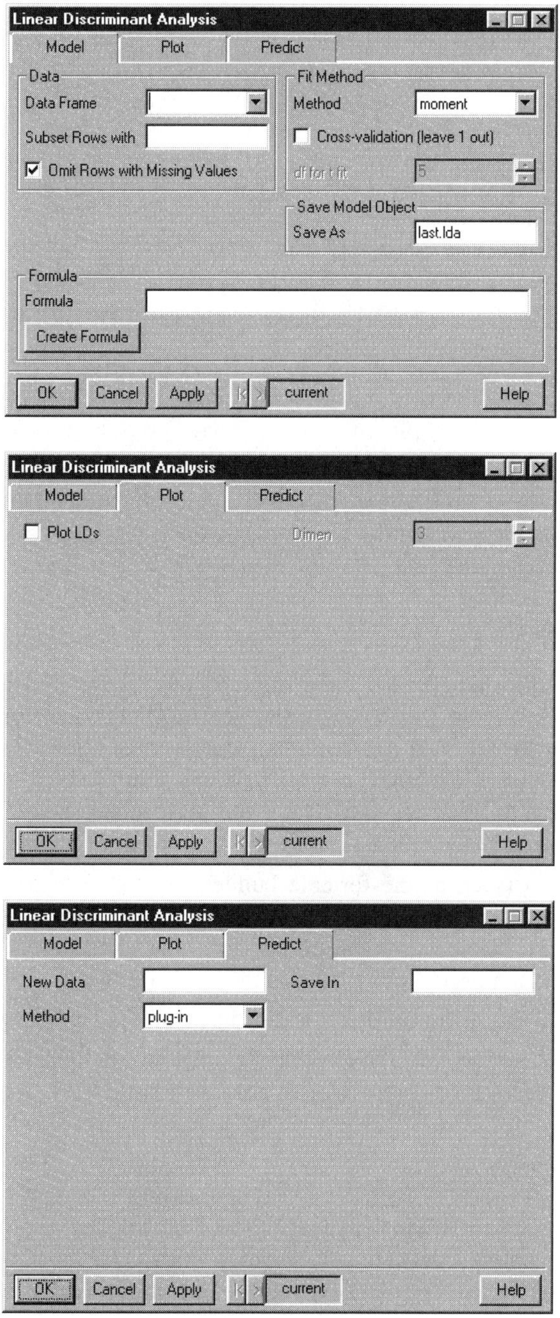

Figure 9.2: Customized dialog box for menuLDA showing each of the three tabbed pages.

```
guiCreate("Property", Name="ldaFitMethod",
          DialogPrompt="Method",
          DialogControl="List Box",
          OptionList="moment, mle, mve, t",
          DefaultValue="moment", UseQuotes=T)
guiCreate("Property", Name="ldaCV",
          DialogPrompt="Cross-validation (leave 1 out)",
          DialogControl="Check Box", DefaultValue=F)
guiCreate("Property", Name="ldaFitNu",
          DialogPrompt="df for t fit",
          DialogControl="Integer Spinner", Range=2:50,
          DefaultValue=5, Disable=T)
```

As the other two pages show in the figure, basic property objects can be used directly on a page. The use of pages is optional: if none are specified all the groups or basic objects are placed on a single page (as in Figures 9.1 and 9.4).

We can begin to see how the argument list of the `FunctionInfo` property is made up of the values of (most of) the fields in the basic property items. Item #0 is used to assign the result of the function call, and #1, #2, ... specify the order in which the fields in the dialog box are matched to the function arguments.

*Dialog box layout*

The layout of the dialog box is done automatically, and optimizing it seems a matter of trial-and-error. Two columns are used, and the groups or items are used in order filling the left then right column. However, as Figure 9.2 shows, both columns will be used, so if there is more than one group or item, at least the last will be placed in the right column. The width of the dialog box is fixed, but the height is adjusted to accommodate all the entries (on all the pages).

Wide groups (such as the formula builder `SPropPFFormulaG`) span both columns, and it may be necessary to add newline properties (for example, `SPropFSpace1`) judiciously to ensure that the wide groups are assigned to the first column and not overlaid by entries in the second column.

The reason why there are functionally identical GUI property objects like `SPropFSpace1` and `SPropFSpace2` is that a GUI object may appear only once in each dialog box.

## Callback functions

There is one argument in the definition of the `FunctionInfo` property that we have not yet mentioned, the *callback function*. This is an S function that is called whenever a button is clicked or a field in the dialog box is altered, as well as when the dialog box is first invoked. It will normally contain calls to the GUI to change other properties of the dialog box. The state information is passed around in a data frame which is most easily manipulated by system functions whose names start with `cb`. For `menu.lda` we used

# *Discover the Power of* (S-PLUS)®

*With S-PLUS, powerful analysis and visualization techniques are combined with an intuitive, customizable interface, so you can explore data with ease.*

## Insightful Analysis & Publication Quality Graphics

Reveal hidden patterns in your data with powerful S-PLUS graphing techniques. Create stunning graphics with point-and-click ease. Control every detail of your graphs and easily produce publication-quality output.

## Cutting-Edge Functionality

Over 3800 built-in functions are available through menus and dialogs, offering unparalleled power for analyzing your data.

## The Power of S

The rich object-oriented S language gives you the power and flexibility to modify or create your own functions.

## Fully Customizable User Interface

Change toolbars, menus and dialogs to suit your working style.

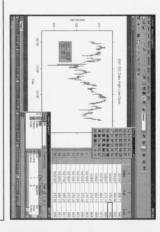

# BUSINESS REPLY MAIL

FIRST-CLASS MAIL PERMIT NO. 75807 SEATTLE, WA

POSTAGE WILL BE PAID BY ADDRESSEE

DATA ANALYSIS PRODUCTS DIVISION
MATHSOFT INC
1700 WESTLAKE AVE N STE 500
SEATTLE WA 98109-9891

```
backLDA <- function(df)
{
  activeprop <- cbGetActiveProp(df)
  activevalue <- cbGetCurrValue(df, activeprop)
  switch(activeprop,
    ldaPlot = {
        df <- cbSetEnableFlag(df, "ldaPlotDimen", activevalue=="T")
    },
    ldaFitMethod = {
      if(activevalue == "moment" || activevalue == "mle") {
        df <- cbSetEnableFlag(df, "ldaCV", T)
      } else {
        df <- cbSetEnableFlag(df, "ldaCV", F)
        df <- cbSetCurrValue(df, "ldaCV", "F")
      }
      df <- cbSetEnableFlag(df, "ldaFitNu", activevalue=="t")
    },
    ldaCV = {
      set <- activevalue == "F"
      df <- cbSetEnableFlag(df, "ldaPlot", set)
      df <- cbSetEnableFlag(df, "ldaPredictMethod", set)
      df <- cbSetEnableFlag(df, "SPropPredictNewdf", set)
      df <- cbSetEnableFlag(df, "SPropSavePredictObject", set)
    },
    SPropPFButton = {df <- backform1(df)},
    SPropPFEnableButton = {
        df <- cbSetEnableFlag(df, "SPropDataFrameList", T)
        df <- cbSetEnableFlag(df, "SPropPFFormula", T)
        df <- cbSetEnableFlag(df, "SPropPFButton", T)
    }
  )
  df
}
```

One purpose of this code is to enable or disable certain fields depending on the settings of others. First we find out what changed and what its new value is. If plotting has been turned on or off, we enable or disable the plot dimension field, and so on. The precise meaning of these function calls can be looked up in the on-line help, but it is normally sufficient to copy examples such as these.

The other purpose is to handle the Create Formula button (by bringing up another dialog box and substituting its result in the appropriate field). This is common code shared by several callback functions, adapted from the callback function backLm for menuLm.

```
backform1 <- function(df)
{
  if(exists(cbGetCurrValue(df, "SPropDataFrameList"))) {
    guiModify("Property", Name = "SPropCFFormulaID",
            DialogControl = "Invisible",
            DefaultValue = cbGetDialogId(df))
```

```
      guiModify("Property", Name = "SPropCFData",
              DialogControl = "Invisible",
         DefaultValue = cbGetCurrValue(df, "SPropDataFrameList"))
      guiModify("Property", Name = "SPropCFFormula",
         DefaultValue = cbGetCurrValue(df, "SPropPFFormula"))
      guiDisplayDialog("Function", Name = "makeFormulaLm")
      df <- cbSetEnableFlag(df, "SPropDataFrameList", F)
      df <- cbSetEnableFlag(df, "SPropPFFormula", F)
      df <- cbSetEnableFlag(df, "SPropPFButton", F)
    } else guiCreate("MessageBox", String = "Data Frame doesn't
            exist. Enter Data Frame before building formula.")
  df
}
```

We have not seen guiModify before: it acts like guiCreate but on an existing property, retaining the values of all the fields that are not specified.

The function guiDisplayDialog is used to display the dialog box for the formula builder. We have followed the system example here, but it might be more natural to set the argument bModal to true (the default is false). In Microsoft-speak a 'modal' dialog box grabs attention and must be dealt with before anything else in the S-PLUS process. An example of a 'modal' dialog box is the message box that we use in the event of an error. The default, a 'modeless' dialog box, allows focus to be transferred to other elements of the GUI. One visible difference is that a 'modeless' dialog box has an Apply button, whereas a 'modal' dialog does not.

### Initialization

The callback function is also called when the dialog box is first launched. Implicitly we have relied on that to help set the initial state of the dialog box. If additional initialization is required, the function IsInitDialogMessage with argument the state-information data frame will return true on the initial call, and false subsequently. This could be used to, for example, customize the dialog box title.

### Validation of entries

Another important application of the callback function is to validate the entries in the dialog box. Although properties such as Integer Spinner can be given a Range argument, that only limits the values that can be selected via the up/down buttons: any value can be typed in.

The operation of validation routines in a callback can be slightly disconcerting: the callback function is called slightly after the pointer is moved away from the field, so the user may perceive the system as having accepted the value and then deciding it does not like it.

As the user is not obliged to change the field value, the callback function needs to, for example by clearing the field (use guiModify to set DefaultValue to "") or changing it to the nearest valid value. See our example on page 217 for one way to handle this.

*Completeness*

We will not want the underlying function, here `menuLDA`, to be called unless all the arguments which are needed have been specified. We can ensure this in one of two ways. Individual fields in the dialog box have the property of `IsRequired`, by default false. If this is specified as true using neither `Apply` nor `OK` will call the function until the field is non-empty.

If the condition for the dialog box to be complete is more complicated than requiring specific fields to be non-empty (for example we may require at least one of a set of fields to be completed) this can be handled by the callback function (un)setting the value of an invisible field which is required. Such a field can be created by

```
guiCreate("Property", Name="IsComplete",
          DialogControl="Invisible", IsRequired=T)
```

Since this is not visible, a user cannot change its value, but the callback function can change the default value. If this field is empty the user will see a message box saying it is required, so it is important to give it an informative name.

## Finding out about available GUI objects

How do we find the details of the GUI objects and properties? The documentation is sparse, and we can either copy existing examples or examine the GUI dialog boxes used to create new properties. In either case an object browser (S-PLUS 4.0, 4.5) or object explorer (S-PLUS 2000) must be used. Create a page with filtering set to the interface class `Property`. There are very many properties, and the right-hand panel will (probably) have many pages. Selecting any property, right-clicking and selecting **Create Property...** from the menu will bring up a dialog box (Figure 9.3) that describes the possible arguments to `guiCreate` for a property: using the **Help** button of that dialog box can be informative. Such dialog boxes can be used to create all the properties: the history log will then show the commands that can be used from S to recreate them. As there is no simple way to find out at a later date what commands have been used to create the properties (or even what properties have been added), it is important to keep a copy of the relevant parts of the history log. However, a command to re-create[2] a property can be generated by dragging the icon for a property from an object explorer/browser to an open *script* window.

As from S-PLUS 4.5 the same information can be retrieved from the command line. Function `guiPrintClass` gives a list of the arguments for `guiCreate` for an object of that class. For example

```
> guiPrintClass("FunctionInfo")
CLASS:   FunctionInfo
ARGUMENTS:
     Name
```

---

[2]including all the fields which were not specified.

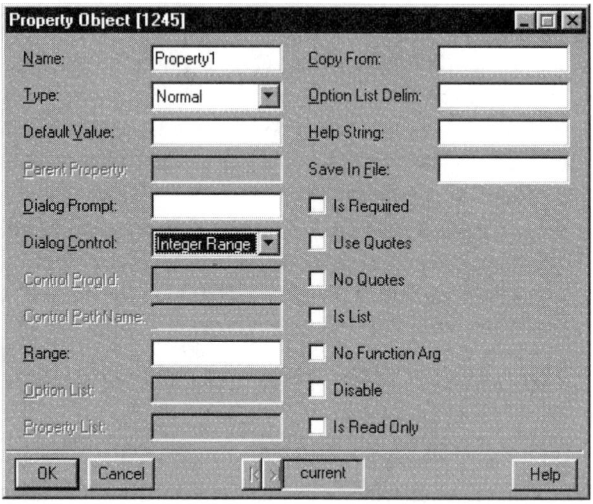

**Figure 9.3**: The dialog box for creating properties, here set to Integer Range. Which fields are enabled depends on the setting of the Dialog Control drop-down selection field.

```
        Prompt:
        Default:
Function
        Prompt:     Function
        Default:    "FunctionInfo1"
DialogHeader
        Prompt:     Dialog Header
        Default:    ""
....
Display
        Prompt:     Display
        Default:    FALSE
        Option List: [ T, F ]
```

If we have more idea of what we are looking for we can use more specific functions:

```
> guiGetArgumentNames("FunctionInfo")
 [1] "Name"              "Function"           "DialogHeader"
 [4] "StatusString"      "PropertyList"       "ArgumentList"
 [7] "ArgumentClassList" "PromptList"         "DefaultValueList"
[10] "Yes"               "CallBackFunction"   "HelpCommand"
[13] "SavePathName"      "WriteArgNames"      "Display"
> guiGetPropertyOptions("FunctionInfo", "Display")
[1] "T" "F"
> guiGetPropertyPrompt("FunctionInfo", "Display")
$PropName:  [1] "Display"
$prompt:    [1] "Display"
```

```
$default:    [1] F
$optional:   [1] T
$data.mode:  [1] "logical"
```

It is possible to retrieve the current values of all of a property object or just parts of it by

```
> nm <- guiGetArgumentNames("Property")
> val <- guiGetPropertyValue("Property", Name="ldaModelPage")
> cbind(Name=nm, Value=substring(val[-(1:2)],1,20))
                        Name                        Value
 [1,] "Name"                    ""
 [2,] "Type"                    "Page"
 [3,] "DefaultValue"            ""
 [4,] "ParentProperty"          ""
 [5,] "DialogPrompt"            "Model"
 [6,] "DialogControl"           "Page Tab"
 [7,] "ControlProgId"           ""
 [8,] "ControlServerPathName"   ""
 [9,] "Range"                   ""
[10,] "OptionList"              ""
[11,] "PropertyList"            "ldaModelData, SPropF"
[12,] "CopyFrom"                ""
[13,] "OptionListDelimiter"     ""
[14,] "HelpString"              ""
[15,] "SavePathName"            ""
[16,] "IsRequired"              "F"
[17,] "UseQuotes"               "F"
[18,] "NoQuotes"                "F"
[19,] "IsList"                  "F"
[20,] "NoFunctionArg"           "F"
[21,] "Disable"                 "F"
[22,] "IsReadOnly"              "F"

> guiGetPropertyValue("Property", Name="ldaModelPage",
                      PropName="DialogPrompt")
[1] "Model"
```

Unfortunately, however it is retrieved the PropertyList will be truncated.

Moving in the other direction, there does not appear to be a comprehensive list of classes. The list includes

```
Application              AutomationClient      ClassInfo
FunctionInfo             GraphSheet            MenuItem
MessageBox               ObjectBrowser         ObjectDefault
OptionsDocumentSettings  OptionsCommandLine    OptionsUndo
OptionsResizeGraph       OptionsCycle          OptionsCycleBW
OptionsColorSchemes      Property              Report
Script                   SearchPath            Toolbar
```

plus the usual classes of S-PLUS objects viewed in the object explorer/browser (and many other classes).

## An example

We will consider an example from Chapter 12 of MASS throughout the rest of this
chapter. We considered the dataset from a Veteran's Administration lung cancer
trial used by Kalbfleisch & Prentice (1980) and provided in dataset `cancer.vet`.
Our final model was `VA.cox3`, and we consider the task of predicting the survival
of a similar patient from that model. To do so we can re-create the model and use

```
VA.temp <- as.data.frame(cancer.vet)
dimnames(VA.temp)[[2]] <- c("treat", "cell", "stime",
    "status", "Karn", "diag.time","age","therapy")
attach(VA.temp)
VA <- data.frame(stime, status, treat=factor(treat), age,
    Karn, diag.time, cell=factor(cell), prior=factor(therapy))
detach()
VA$Karnc <- VA$Karn - 50
options(contrasts=c("contr.treatment", "contr.poly"))
VA.cox <- coxph(Surv(stime, status) ~ treat + age  + Karn +
                diag.time + cell + prior, VA)
VA.cox3 <- update(VA.cox, ~ treat/Karnc + prior*Karnc
                + treat:prior + cell/diag.time)

menuVA <- function(treat, age, Karn, diag.time, cell, prior)
{
  treat <- factor(match(treat, c("standard","test")),
                  levels=1:2, labels=levels(VA$treat))
  cell <-
    factor(match(cell, c("squamous","small","adeno","large")),
           levels=1:4, labels=levels(VA$cell))
  prior <- factor(match(prior, c("no", "yes")), levels=1:2,
                  labels=levels(VA$prior))
  nd <- data.frame(treat, age, Karnc=Karn-50, diag.time,
                   cell, prior)
  plot(survfit(VA.cox3, nd, conf.type="log-log"))
  invisible()
}
```

which given a set of values plots the predicted survival curve with a confidence
interval. In validating the values we need to allow the categorical variables to take
the values they are matched against, Karnofsky score is in $[0, 100]$ and diagnosis
time is positive (in months).

The custom dialog box (Figure 9.4) is defined by

```
guiCreate("FunctionInfo", Name = "menuVA",
    Function = "menuVA",
    HelpCommand = "help(VA)",  # which may not exist!
    DialogHeader = "VA Cancer Prediction",
    PropertyList = c("SPropInvisibleReturnObject","VAtreat",
        "VAage", "VAKarn", "VAdiag", "VAcell", "VAprior"),
    CallBackFunction = "backVA",
```

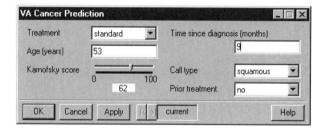

**Figure 9.4**: Dialog box for menuVA .

```
ArgumentList = "#0=SPropInvisibleReturnObject, #1=VAtreat,
    #2=VAage, #3=VAKarn, #4=VAdiag, #5=VAcell, #6=VAprior")
```

```
guiCreate("Property", Name="VAtreat", DialogPrompt="Treatment",
        DialogControl="List Box", OptionList="standard, test",
        DefaultValue="standard", UseQuotes=T)
guiCreate("Property", Name="VAcell", DialogPrompt="Call type",
        DialogControl="List Box",
        OptionList="squamous, small, adeno, large",
        DefaultValue="squamous", UseQuotes=T)
guiCreate("Property", Name="VAprior",
        DialogPrompt="Prior treatment",
        DialogControl="List Box", OptionList="no, yes",
        DefaultValue="no", UseQuotes=T)
guiCreate("Property", Name="VAage",
        DialogPrompt="Age (years)",
        DialogControl="Float",
        DefaultValue="", IsRequired=T)
guiCreate("Property", Name="VAKarn",
        DialogPrompt="Karnofsky score",
        DialogControl="Float Slider", Range=0:100,
        DefaultValue="", IsRequired=T)
guiCreate("Property", Name="VAdiag",
        DialogPrompt="Time since diagnosis (months)",
        DialogControl="Float",
        DefaultValue="", IsRequired=T)
```

and the callback function is

```
backVA <- function(df)
{
  activeprop <- cbGetActiveProp(df)
  activevalue <- cbGetCurrValue(df, activeprop)
  switch(activeprop,
    VAKarn = {
      activevalue <- as.numeric(activevalue)
      if(activevalue < 0) {
        df <-  cbSetCurrValue(df, "VAKarn", 0)
```

```
                  guiCreate("MessageBox",
                            String="0 <= Karnofsky score <= 100")
          }
          if(activevalue > 100) {
            df <- cbSetCurrValue(df, "VAKarn", 100)
            guiCreate("MessageBox",
                      String="0 <= Karnofsky score <= 100")
          }
        },
        VAage = {
          activevalue <- as.numeric(activevalue)
          if(activevalue < 0) {
            df <- cbSetCurrValue(df, "VAage", 0)
            guiCreate("MessageBox", String="Age must be positive")
          }},
        VAdiag = {
          activevalue <- as.numeric(activevalue)
          if(activevalue < 0) {
            guiCreate("MessageBox",
                      String="Time to diagnosis must be positive")
            df <- cbSetCurrValue(df, "VAdiag", 0)
          }}
          )
      df
}
```

## 9.2   Adding items to the menus

Menu items are added in the same way as dialog boxes, by adding GUI properties
this time of class `MenuItem`. To add a separate menu for the MASS library we
could use

```
stat.loc <- guiGetPropertyValue("MenuItem", Name = paste(
            guiGetMenuBar(), "Statistics", sep = "$"),
            PropName = "Index")
MASSmenu <- paste(guiGetMenuBar(), "MASS", sep = "$")
guiCreate("MenuItem",
        Name="MASSmenu", Type="Menu",
        MenuItemText= "&MASS", Index=stat.loc+1, OverWrite=F)
guiCreate("MenuItem",
        Name=paste(MASSmenu, "LDA", sep="$"), Type="MenuItem",
        Action="Function", Command="menuLDA",
        MenuItemText= "&LDA...")
    ....
# to remove this use
guiRemove("MenuItem", Name=MASSmenu)
```

The numbering of the menus is potentially confusing, as not all menus are enabled at any one time. The way to discover the numbers required is to open an object explorer/browser, filter on the interface class MenuItem and explore the full menu tree. This example adds a MASS menu after the Statistics menu (which was number 9 in S-PLUS 4.5). As the menu structure can be re-programmed by the user, in general one can assume nothing about the existing menus. (They could be in a different human language, for example.) The code given will calculate the right menu position, but only in S-PLUS 2000.

Menus have a tree structure, and an item on a menu is specified by giving the path from the root to that item. This is very like the specification of URLs except that the path separator is $; the root of the menu tree is written as $$ when needed.

The & in the menu item text precedes a letter that will be underlined and can be used as a short cut to the item: it should be unique within the current level of the menu hierarchy.

## Other menu items

So far we have seen the creation of menu item objects of types Menu and MenuItem; the third possibility is Separator that creates the horizontal lines in menus separating groups of menu items.

A menu item does not necessarily call up a dialog box; the possibilities are

None Do nothing. This can be useful to leave a stub for planned extensions.

Builtin Use any of the standard operations in the GUI normally associated with menu items or buttons on toolbars or palettes. What is available will be system-dependent and is best explored from the dialog box for a new menu item in an object explorer or object browser.

Open Open a file, using the application associated with that file type in Windows.

Print Print a file.

Run Open a file as a script window, and run the contents of the script.

Expression Evaluate an S expression.

Function Open the dialog box for an function; this can be by-passed and the function called with its default arguments by setting ShowDialogOnRun=F.

For Open, Print and Run the file name is specified as the Command argument to guiCreate.

## Context-sensitive menus

Another use for menus is to provide class-specific items on the pop-up menu produced by right-clicking (or double-clicking) on an item in an object explorer/browser. First we define the additional items for this menu and then assign it (this time as a GUI object of type `ClassInfo`).

```
guiCreate("MenuItem", Name="lda", Type="Menu",
          DocumentType="lda")
guiCreate("MenuItem", Name="lda$print", Type="MenuItem",
          DocumentType="lda", Action="Function",
          Command="tabResults.LDA", MenuItemText="Print",
          ShowDialogOnRun=F)
guiCreate("MenuItem", Name="lda$plot", Type="MenuItem",
          DocumentType="lda", Action="Function",
          Command="tabPlot.LDA", MenuItemText="Plot...",
          ShowDialogOnRun=T)
guiCreate("MenuItem", Name="lda$predict", Type="MenuItem",
          DocumentType="lda", Action="Function",
          Command="tabPredict.LDA", MenuItemText="Predict...",
          ShowDialogOnRun=T)

guiCreate("ClassInfo", Name="lda", ContextMenu="lda",
          DoubleClickAction="tabResults.LDA")
```

We reuse the S functions that we have already created for use in `menuLDA`.

However, these functions now need to be associated with dialogs, and the appropriate defaults will be slightly different from those of the main dialog box. Some appropriate definitions are

```
guiCreate("FunctionInfo", Name = "tabPlot.LDA",
          Function = "tabPlot.LDA",
          DialogHeader = "LDA plot",
          PropertyList = c("SPropInvisibleReturnObject",
            "SPropCurrentObject","ldaPlot2Page",
            "SPropPFEnableButton"),
          ArgumentList = "#0=SPropInvisibleReturnObject,
                   #1=SPropCurrentObject, #2=ldaPlotDimen2")
guiCreate("Property", Name = "ldaPlot2Page", Type = "Page",
          DialogPrompt = "Plot",
          PropertyList = "ldaPlotDimen2")
guiCreate("Property", Name = "ldaPlotDimen2",
          DialogPrompt = "Dimen",
          DialogControl = "Integer Spinner", DefaultValue = 3)

guiCreate("FunctionInfo", Name = "tabPredict.LDA",
          Function = "tabPredict.LDA",
          DialogHeader = "LDA prediction",
          PropertyList = c("SPropInvisibleReturnObject",
          "SPropCurrentObject","ldaPredictPage",
```

```
                  "SPropPFEnableButton"),
          ArgumentList = "#0=SPropInvisibleReturnObject,
            #1=SPropCurrentObject, #2=SPropPredictNewdata,
            #3=SPropSavePredictObject, #4=ldaPredictMethod")
```

Note the use of SPropInvisibleReturnObject; this discards the result of the function, and does not create a visible field in the dialog box to match the item.

## 9.3   Managing a customized GUI

As the GUI properties are stored in the user's preferences files, they are not easily shared in a library, for example. They can be added and removed by S-PLUS functions that call guiCreate and guiRemove. For example, our library MASSdia has functions

```
addMASSmenus()
removeMASSmenus()
```

to add and remove the GUI objects it needs, and users are asked to use these by a message printed by the .First.lib function of the library. These functions could be invoked in .First.lib and .Last.lib respectively, but they are rather slow and we have found this process to be error-prone. Given the slowness, we prefer to load up the GUI objects once per project.

As S-PLUS can be run without the GUI (see Appendix C), it is not a good idea to create or manipulate GUI objects from .First. From S-PLUS 4.5 a .guiFirst function can be used: this is only invoked if the GUI is present. As a sanity check, addMASSmenus begins

```
addMASSmenus <- function()
{
   if(!exists("guiCreate")) stop("Not running under the GUI")
      ....
```

Without this, the user might have to acknowledge hundreds of alert boxes.

It is possible to share _Prefs directories between projects by using the S_PREFS environmental variable in the shortcut or in the environment. Note that the whole path, including _Prefs, should be specified.

The functions guiLoadDefaultObjects and guiStoreDefaultObjects can be used to load or store from a file other than the standard files. The SavePathName property of an GUI object specifies where it will be saved: by default the file from which it was loaded.

## Restoring the defaults

In experimenting with menus and properties it is easy to leave unwanted elements behind, and on occasion to remove wanted properties or parts of the menu tree. It is helpful to know that the menu tree is stored in the file _Prefs\smenu.smn and the property objects in _Prefs\axprop.dft; if either of these files is deleted it will be replaced by a copy of the system version of the file when S-PLUS is next started. Similarly, the FunctionInfo and ClassInfo objects are stored in files _Prefs\axfunc.dft and _Prefs\axclinfo.dft.

## Using the built-in GUI operations

From S-PLUS 4.5 it is possible to examine and set from S code most of the GUI options available on the Options menu. The functions guiGetOption and guiSetOption will interrogate and set respectively the value of any option. To find out what the options are called look at the help page for guiSetOption or use

| General Settings | guiPrintClass("OptionsDocumentSettings") |
| Command Line | guiPrintClass("OptionsCommandLine") |
| Undo & History | guiPrintClass("OptionsUndo") |
| Graph Options | guiPrintClass("OptionsResizeGraph") |
| Graph Styles \| Color | guiPrintClass("OptionsCycle") |
| Graph Styles \| Black and White | guiPrintClass("OptionsCycleBW") |
| Color Schemes | guiPrintClass("OptionsColorSchemes") |

to print the options for the appropriate menu item.

The function guiExecuteBuiltIn can be used to select any builtin action from the menu tree. For example

```
guiExecuteBuiltIn("$$SPlusMenuBar$No_Documents$Help$
            Search_Language_Reference", wait=F)
```

(all one line) can be used to bring up the index tab of the S-PLUS help file, and not wait for the user to quit the help viewer (waiting being the default for guiExecuteBuiltIn). To find out the appropriate name, see the list of possibilities for the Builtin action of a menu item, given by

```
guiGetPropertyOptions("MenuItem", "BuiltInOperation")
```

At the time of writing there were 1021 items.

Other options settings can be retrieved or set by other functions, for example setTextOutputRouting.

## 9.4 Communicating with S-PLUS: DDE

Another way to provide a user interface is to use S-PLUS as the 'engine' to run computations, driven by an external Windows program more familiar to the users, for example a spreadsheet or a custom-made application written in a visual programming language. There are two main ways to do this, *Dynamic Data Exchange*, and *Automation* which we cover in the next section. Automation is more comprehensive but requires a higher level of support from the client program.

S-PLUS can act as a DDE server, and will do so unless disabled (in the General Settings dialog box on the Options menu). Dynamic Data Exchange allows an external program to start a conversation with S-PLUS, submit S code to S-PLUS, and optionally receive back the results as character strings. The effect is as if the external program was typing commands in the commands window, and perhaps grabbing the output. DDE is a two-way process, but S-PLUS cannot use it to initiate communications: it is a DDE server but not a DDE client.[3]

Many standard Windows office applications can act as DDE clients, and DDE clients can be programmed in any programming language with a full-featured Windows compiler: Visual Basic is particularly popular and we have used Visual C++ and gcc.

S-PLUS must already be running[4] to communicate using DDE. The server name for S-PLUS is S-PLUS. One type of DDE session might be caricatured as

> Initiate the conversation by contacting the DDE server S-PLUS with the DDE topic SCommand. This will return a DDE channel number to be stored by the client. (If this is unsuccessful, the client could attempt to launch S-PLUS.)
>
> Send commands to S-PLUS with the DDE topic SCommand or (equivalently) [Execute] followed by S code. The code is executed as if typed in the commands window, and the results appear there (or wherever text routing has been selected to appear).
>
> Send commands to S-PLUS with the DDE topic [Request]. Wait to be notified that the output data is ready, allocate storage for it and ask for it to be sent to the client.
>
> Shut down the conversation.

Higher-level facilities in the client language may abstract the details: for example in Visual Basic for Applications (for example, from Excel) the conversation

```
Channel = Application.DDEInitiate("S-PLUS", "SCommand")
ObjectList = Application.DDERequest(Channel, "ls()")
Application.DDETerminate(Channel)
```

handles the waiting for the S function to complete, the allocation of storage and the transfer of the results.

---

[3]except for DDE linking of blocks of cells in S-PLUS data objects.

[4]if more than one copy is running, the results are unpredictable.

Another type of DDE conversation involves interrogating part of a data frame, along the lines of

Initiate the conversation by contacting the DDE server S-PLUS with the DDE topic the name of the data frame. This will return a DDE channel number to be stored by the client.

Request data with the DDE topic [Request]: the request is for a rectangular block of cells in spreadsheet notation, for example r1c1:r20c3.

Modify the data with DDE topic [Poke], followed by a cell range and replacement data for the block of cells.

Shut down the conversation.

Simultaneous DDE conversations are supported, so data frames can be interrogated during a commands conversation.

*Using* Visual Basic

A fairly comprehensive DDE test client is supplied in the S-PLUS directory samples\dde\vbclient, and as an executable is supplied, this can be used to test out DDE ideas without having[5] the Visual Basic development system. The source code provides comprehensive examples for Visual Basic programmers. Here are a couple of simpler examples. Suppose we have a text box Text1 and a button Command1. Executing

```
Private Sub Command1_Click()
  Dim sExecute As String
  sExecute = Trim(Text1.Text)
  Text1.LinkMode = vbLinkNone
  Text1.LinkTopic = "S-PLUS|SCommand"
  Text1.LinkItem = ""
  Text1.LinkMode = vbLinkManual
  Text1.LinkExecute sExecute
End Sub
```

by clicking on the button will send the contents of the text box to S-PLUS to be executed (and the result printed there). Alternatively,

```
Private Sub Command1_Click()
  Dim sExecute As String
  sExecute = Trim(Text1.Text)
  Text1.LinkMode = vbLinkNone
  Text1.LinkTopic = "S-PLUS|SCommand"
  Text1.LinkItem = sExecute
  Text1.LinkMode = vbLinkManual
  Text1.LinkRequest
End Sub
```

will send the test to S-PLUS and print the result in the text box.

---

[5] You will need the Visual Basic 4.0 run-time DLL, VB40032.dll, to be installed.

*Using* C

MathSoft do not provide an example of a DDE client in C. We wrote one for use with the ESS package in NTemacs. The file Sdde.c in our on-line files can be compiled under Visual C++ as a console application in the development environment or from the command line by

```
cl Sdde.c /link user32.lib
```

It should also compile under other compilers that have the required header files; for example, with the mingw32 port of gcc 2.95.2 we used

```
gcc -O2 -s -o Sdde Sdde.c
```

This provides a general test client for DDE requests, but with some features tailored to S-PLUS. To use it, run S-PLUS, then run this from a command window by

```
Sdde S-PLUS SCommand
```

A > prompt will appear. Type in S commands: these are sent to the S-PLUS process and the results will be listed in the command window. Quit the client by Ctrl-C. This client can be useful in its own right, but is also intended as an example from which a more sophisticated interface could be built. (Note that the response from S-PLUS is separated by tabs which we convert to newlines.)

## 9.5   Communicating with S-PLUS: Automation

*ActiveX Automation* (formerly known as OLE Automation) is Microsoft-speak for operating system facilities that allows servers such as S-PLUS to 'expose' 'objects' for use in an 'automation' client program.[6] The idea is that the client can manipulate the server's GUI and retrieve results, often through linking or embedding objects provided by the server. The 'objects' are not primarily S objects (although they can be), but objects within the GUI. For the rest of this section 'automation' and 'object' are used in these specialized senses.

The automation server, here S-PLUS, is started if necessary when requested to serve an exposed automation object. Objects must be exposed by registering them: the registration information is stored in the Windows registry with the name of the server (S-PLUS). This means that when S-PLUS is launched automatically it uses the default project directory, so using automation when storing S functions in different projects needs to be handled with care, and it is usual safest to start up the S-PLUS process oneself.

Automation clients can be written in many languages, but by far the most popular seems to be Visual Basic and its variant, Visual Basic for Applications. It is also possible to use C or C++ (with most Windows compilers) or even Perl

---

[6]'controller' in Microsoft-speak.

(Schwartz *et al.*, 1997). S-PLUS can also be used as an 'automation' client, something we will not discuss here.

The installation procedure for S-PLUS normally registers a number of automation objects relating to the S-PLUS GUI and to graphsheets. If this step has been skipped or the registration lost, it can be performed later by running

```
Splus /RegisterOleObjects all
```

or calling `register.all.ole.objects()` from the S-PLUS command line. We will assume that this has been done.

In this section we assume that S-PLUS 4.5 or later is being used: version 4.5 has considerably enhanced facilities for automation. There are several Visual Basic and a few Excel VBA examples in the `samples\oleauto` directory under the S-PLUS main directory. As S-PLUS passes everything as variant types, the interface works well with Visual Basic and Perl, but is painful to handle from C.

Automation can be used in several different ways. We will illustrate it by first using a command-level interaction, to send strings as S commands and receive the results, then to call S functions, set their parameters via the GUI dialog boxes and display the results, and finally to set data as arguments to an S function and receive the result. To complicate the description, results generated by S-PLUS can be displayed by the client, and we do this by linking an S-PLUS graphsheet to a *container* in our client. Although the graph appears to be part of the client, all the plotting is done by S-PLUS.

## Command-line interaction

S-PLUS exposes an `Application` object which represents the current S-PLUS session. Two of its methods, `ExecuteString` and `ExecuteStringResult`, can be used to pass a command to S-PLUS to execute and (for the second) to return the result. These act just like DDE with topics `[Execute]` and `[Request]` respectively, except that automation will start up S-PLUS if required.

We can illustrate this using a Perl program: the precise details will depend on the version of Perl in use.

```
# ActiveState Perl distribution
use OLE;
$cmd = 'ls()'; # just for illustration
$splus = CreateObject OLE "S-PLUS.Application" ||
            die "CreateObject: $!";
$splusver = $splus->Version;
$splused = $splus->Edition;
print "Using S-PLUS $splusver $splused Edition\n";
$splus->ExecuteString($cmd) || die "ExecuteString: $!";
```

This sends a command as a string to S-PLUS; the result is boolean and indicates if the operation succeeded. We also retrieve a couple of properties of the

S-PLUS automation server, to illustrate how this can be done. All automation clients will use a call like `CreateObject` (the same name is used in Visual Basic) to establish the connection, and use the registered `ProgID`, here `"S-PLUS.Application"`, to identify the automation server and object. We will see other objects in due course. Perl uses '`->`' to select methods and properties of automation objects; Visual Basic uses '`.`', so an equivalent snippet is

```
Dim pApp As Object
Set pApp = CreateObject("S-PLUS.Application")
Debug.Print pApp.Version, pApp.Edition
pApp.ExecuteString("ls()")
```

where `Debug.Print` prints the results in the 'immediate' window.

To retrieve the result of the S command we can use

```
use OLE;
$cmd = 'ls()';
$splus = CreateObject OLE "S-PLUS.Application" ||
            die "CreateObject: $!";
$splus->{Visible} = 0;        # hide main S-PLUS window
$spluscmd = $splus->ExecuteStringResult($cmd, 0);
$spluscmd =~ s/\t/\n/g;       # convert tabs to newlines
print "$spluscmd\n";
```

This can return the result in one of two forms. If the second parameter to `ExecuteStringResult` is true, the old (S-PLUS 3.3) form is used. In the current form the lines of the result are (as for DDE) returned separated by tabs, which we convert to newlines (a particularly easy task in Perl). If the command fails, the S error message will be returned (and also sent to the current output text routing in S-PLUS).

### Finding out what you can do with an automation object

There are at least two ways to find out what properties (variables) and methods (functions) are associated with the S-PLUS Application object (or any other automation object). One is to use an OLE browser such as Microsoft's `ole2vw32.exe`[7] to browse the S-PLUS type library `sp4obj.tlb` in the `cmd` directory under the S-PLUS home directory. Users of Visual Basic can use its internal object browser after adding to the project a *reference* to that file as the S-PLUS 4.0/2000 Object Library.

The other way is to ask the server directly using automation:

```
use OLE;
$splus = CreateObject OLE "S-PLUS.Application" ||
            die "CreateObject: $!";
$spProp = $splus->Properties(); $spMeth = $splus->Methods();
print "Properties: $spProp\nMethods: $spMeth\n";
```

---

[7]provided with several compilers.

which will print two comma-delimited lists of properties and methods. The Visual Basic equivalent is

```
Dim pApp As Object
Set pApp = CreateObject("S-PLUS.Application")
Debug.Print pApp.Properties, pApp.Methods
```

This route is the only way to explore dynamically created child objects (such as the linked graphsheets that we will use later on).

## Interaction via the S-PLUS GUI

Another way to interact with S-PLUS using automation is to launch the dialog box for a function, and let the user interact with that. We will try that for our cancer prediction function menuVA. The first thing we have to do is to 'expose' menuVA by registering it:

```
register.ole.object("menuVA")
```

We can then launch its dialog box using one of the methods ShowDialog, ShowDialogInParent and ShowDialogInParentModeless. The first shows the dialog box as a window within the S-PLUS window (as happens when the dialog box is launched from within S-PLUS). The other two show the dialog box within a specified window; ShowDialogInParent generates a dialog without an Apply button and waits for the function to finish whereas ShowDialogInParentModeless has an Apply button and returns immediately. We can use this in Perl

```
use OLE;
$VA = CreateObject OLE "S-Plus.menuVA" ||
            die "CreateObject: $!";
$VA->ShowDialogInParentModeless(0);
```

(0 is the root window) and in Visual Basic as the event procedure for a button by

```
Private Sub Command1_Click()
    Dim pApp As Object
    Set pApp = CreateObject("S-PLUS.menuVA")
    pApp.ShowDialogInParent(hwnd)
End Sub
```

This code is given in the Visual Basic project VBdialogs1 in the on-line scripts. In a real-life example we might want to ensure that the S-PLUS application is visible before a call to the ShowDialog method, and to detect when the user quits the ShowDialogInParent dialog box.

Automation requests can be batched between calls to the BeginTransaction and CommitTransaction (or CancelTransaction) methods of the S-PLUS objects Application, GraphSheet and of other GUI objects such as graph elements and data windows. Normally these objects update automatically whenever

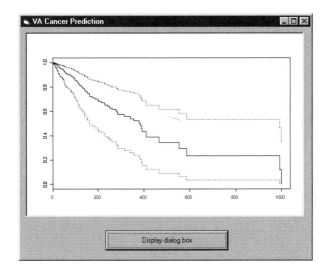

**Figure 9.5**: Interface built in Visual Basic for VA cancer prediction with a linked graphsheet.

one of their properties is changed, so batching changes to properties allows them to be applied simultaneously.

It is possible to link or embed OLE objects such as an S-PLUS graphsheet in a Visual Basic application and display plots on that graphsheet, as shown in Figure 9.5, created by the code in project VBdialogs2 shown in Figure 9.6.

The large window was created as an OLE container, with an S-PLUS GraphSheet linked to it. The Form_Load procedure finds the path name of this graphsheet (it will be something like $$GS14) and uses automation to call the S function graphsheet with argument Name equal to the name of the embedded graphsheet. (This sets the current graphics device to that graphsheet.) When the Display Dialog Box button is clicked, the S-PLUS dialog box (Figure 9.4) for menuVA is displayed, modally, and when the user clicks on the OK button of that box, S-PLUS processes the prediction and draws on the current graphsheet (the embedded one). Finally, a call to the CommitTransaction method for the graphsheet flushes the queue of graphics calls, and the plot appears. Clicking the button again will repeat the process.

Note that this process will work whether or not an S-PLUS session is already running, and without any S-PLUS window being visible.

### Interaction directly with S functions

S functions can be exposed and unexposed by S calls like

```
register.ole.object(c("myfun1", "myfun2"))
unregister.ole.object("myfun2")
```

```
Dim pGraphSheet As Object
Dim pApp As Object

Private Sub Dialog__Click()
    Dim pVA As Object
    Set pVA = CreateObject("S-PLUS.menuVA")

    Dim bSuccess As Boolean
    bSuccess = pVA.ShowDialogInParent(hWnd)
    If bSuccess Then
        Set pGraphSheet = OLE1.object
        pGraphSheet.CommitTransaction
    End If
End Sub

Private Sub Form_Load()
    Dim gname As String
    Set pGraphSheet = OLE1.object
    gname = Mid(pGraphSheet.PathName, 3)
    Set pApp = CreateObject("S-PLUS.Application")
    pApp.ExecuteString "graphsheet(Name='" + gname + "')"
End Sub
```

**Figure 9.6**: Source code for the Visual Basic project VBdialogs2.

As a trivial example, consider the function

```
OLEtest <- function(a, b) a*b
register.ole.object("OLEtest")
```

This registers an object with properties ReturnValue, a and b. As the return value will be a vector, it is simpler to use Visual Basic than Perl. For example, we can use

```
Private Sub Command1_Click()
    Dim pOLEtest as Object
    Dim pReturnArray as Variant

    Set pOLEtest = CreateObject("S-PLUS.OLEtest")
    pOLEtest.SetParameterClasses("vector, vector, vector")
    pOLEtest.a = 1.5
    pOLEtest.b = 2.7
    pOLEtest.Run
    pReturnArray = pOLEtest.ReturnValue
    Debug.Print pReturnArray(1,1)
End Sub
```

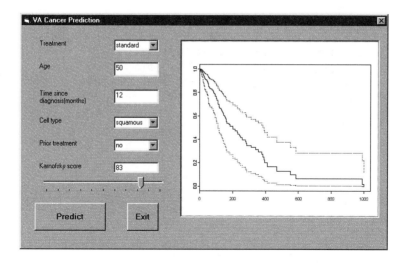

**Figure 9.7**: Interface built in Visual Basic for menuVA.

as the click event procedure for a button, which will call our function OLEtest in S-PLUS and report the result in the immediate window. There is no need to batch the setting of properties here, as for S-PLUS functions no action is taken until the Run method is called.

The SetParameterClasses call is explained on page 233.

We can use the same principles to build a visual front-end to our function menuVA. Make a copy[8] of menuVA called mVA, and register it by

```
register.ole.object("mVA")
```

The registered name of the argument diag.time becomes diagtime, as variable names in Visual Basic (for example) cannot contain periods.

Figure 9.7 shows the interface we constructed in Visual Basic, in project VBdialogs3. When the Predict button is clicked this validates the entries, and calls mVA in S-PLUS, which then displays the plot of the predicted survival distribution on the current device. (This example could be elaborated in many ways, but our aim is to explore communication with S-PLUS, not Visual Basic programming!)

The Visual Basic code used was

```
Dim pGraphSheet As Object
Dim pApp As Object

Private Sub Form_Load()
```

---

[8]so it does not have a FunctionInfo object associated with it. If it does, the arguments have to be given the names associated with them in the FunctionInfo object.

```
      Dim gname As String
      Set pGraphSheet = OLE1.object
      gname = Mid(pGraphSheet.PathName, 3)
      Set pApp = CreateObject("S-PLUS.Application")
      pApp.ExecuteString "graphsheet(Name='" + gname + "')"
End Sub

Private Sub Predict_Click()
      Dim pVA As Object
      Dim msg As String
      Dim bSuccess As Boolean

      msg = ""
      If Len(AgeText.Text) = 0 Then
          msg = msg + ", Age"
      End If
      If Len(DiagTimeText.Text) = 0 Then
          msg = msg + ", Diagnosis time"
      End If
      If Len(KarnText.Text) = 0 Then
          msg = msg + ", Karnofsky score"
      End If
      If Len(msg) > 0 Then
          MsgBox ("Incomplete fields:" + Mid(msg, 2))
      Else
          Set pVA = CreateObject("S-PLUS.mVA")
          pVA.SetParameterClasses ("vector, vector, _
            vector, vector, vector, vector, vector")
          pVA.Treat = Treatment.Text
          pVA.Age = CDbl(AgeText.Text)
          pVA.Karn = CDbl(KarnText.Text)
          pVA.DiagTime = CDbl(DiagTimeText.Text)
          pVA.Cell = CellType.Text
          pVA.Prior = Prior.Text
          Form1.MousePointer = 11 ' hourglass cursor
          bSuccess = pVA.Run
          Form1.MousePointer = 0
          If bSuccess Then
              Set pGraphSheet = OLE1.object
              pGraphSheet.CommitTransaction
          End If
      End If
End Sub

Private Sub Exit_Click()
      End
End Sub

Private Sub KarnScore_Scroll()
      KarnText.Text = CStr(KarnScore.Value)
```

```
End Sub

Private Sub KarnText_LostFocus()
    Dim Karn As Integer
    If (Len(KarnText.Text) > 0) Then
        Karn = CInt(KarnText.Text)
        If (Karn < 0 Or Karn > 100) Then
            KarnText.Text = ""
            MsgBox ("0 <= Karn <= 100")
        Else
            KarnScore.Value = Karn
        End If
    End If
End Sub

Private Sub AgeText_LostFocus()
    Dim Age As Integer
    If Len(AgeText.Text) > 0 Then
        Age = CInt(AgeText.Text)
        If Age <= 0 Or Age >= 120 Then
            AgeText.Text = ""
            MsgBox ("0 < Age < 120")
        End If
    End If
End Sub
```

---

This example uses a slider control that may not be available in the 'Learning Edition' of Visual Basic.

Notice that the names of the function arguments are capitalized in the Visual Basic code; case is ignored in matching them to the S-PLUS function's exposed arguments.

It may be useful to bear in mind that all the programming can be done from the client, for the ExecuteString method allows functions to be sent to S-PLUS and exposed as objects, when they can be called directly. Data can be transferred to S-PLUS either as function arguments or by remotely invoking data import functions.

### *Classes of arguments of exposed functions*

Both our examples contain a call to the SetParameterClasses method for the automation object exposed by registering an S function. This is needed as the default 'class' is assumed to be a data frame. The 'class' here is one of the classes of GUI objects that can be displayed in an S-PLUS object explorer / browser, so the likely choices are vector, data.frame, list and matrix. Notice that there is one more class than the number of arguments, as the first item is the class of ReturnValue.

An alternative way to specify the classes of the return value and arguments is to do so on the S-PLUS side as the `ArgumentClassList` argument by using `guiCreate` to set the `FunctionInfo` object for the function. Unfortunately, using a `FunctionInfo` object changes the registered names for the parameters to those specified as names for the dialog box fields, so considerable care is needed. For mVA we could use

```
guiCreate(
  "FunctionInfo", Function = "mVA",
  ArgumentClassList = paste(rep("vector", ), sep=", "))
```

## 9.6   Interfacing with R

The Windows version of R is built as a DLL with several front ends, including the usual GUI and terminal interfaces. One example front-end gives a direct connection to the R command line: the user supplies a function that is called to feed R with commands, and another function for R to call to 'print' the results. Graphics windows and pagers are launched in the usual way. This approach allows a customized front end to be built in C or C++, or via a C wrapper, any language that can interface to DLLs.

Another approach is to build a DDE or automation server as a front end, and an interface to Excel has been built that way. Using DCOM the automation server and client can be on different machines, even in different countries. An example DCOM server by Thomas Baier which can be used with both Visual Basic and C is available from the R distribution sites.

## 9.7   Some pitfalls of Automation

It needs to be borne in mind that a single copy of S-PLUS or R is providing DDE or automation services, potentially to many clients. As (at least at present) the servers are single-threaded, the requests will be dealt with in turn (probably as a queue in order of submission) but it is quite possible for clients to interfere with each other, for example by manipulating S objects of the same name. Sending an incomplete command expecting to be able to complete it at the next send is certainly unwise.

More generally, automation is designed with stateless systems in mind. We have already commented that a currently running instance of the server will be used, with whatever working directory it happens to have. Careful implementation is needed.

# Appendix A

# Compiling and Loading Code

S-PLUS and R already have many compiled FORTRAN and C functions loaded for use by functions such as svd, qr and eigen. There is a confusing variety of methods for loading other functions, such as those from our convolve.c (page 126), but only a few of these are available for each platform. We will assume that suitable compilers are available: please consult your system manuals for details of what is needed.

Note that with some of these methods it is necessary to arrange that your code is loaded by an explicit call to dyn.load, dyn.load.shared or dll.load and for some loading is automatic.

## A.1   Procedures with S-PLUS

### S-PLUS 3.x on UNIX

This system has the widest range of possible methods, but no platform has all of them available.

1. *Static loading* (all platforms).   This option creates a private copy of the S-PLUS executable for the user which includes all the S-PLUS routines together with any others that may be needed. Since the executable is fairly large this should only be done in special circumstances, such as when a group of users will use the same copy, or when the alternative facilities are not applicable.

   There is a LOAD facility. For our function it amounts to running S-PLUS with an initial command line argument

   ```
   $ Splus LOAD convolve.c
   ```

   This does several things. First it compiles the external routines using the compiler appropriate to the file extension. Then it creates an executable file called local.Sqpe in the local directory including the extra routines. Once this executable is made, when S-PLUS is invoked from that directory it uses this file instead of the system executable.

This method is the least desirable, but sometimes is the only feasible route. Libraries for both C and FORTRAN are needed, which may mean that both compilers need to be available, and it can be important to use precisely the right version of the compilers to ensure that compatible libraries are used. (Using the wrong versions can have unpredictable consequences on existing code in S-PLUS, not just your own extensions.) We have always managed to avoid this solution.

2. *Incremental loading* with dyn.load (SunOS4, Sun Solaris 2.x and IBM RS/6000). This is possibly the most convenient where available. The dyn.load function dynamically loads object files and makes the functions available to the version of S-PLUS presently running. Each file should be compiled in the normal way (but see below). If more than one file is to be loaded, they can be accumulated into a single relocatable object file using

```
$ ld -r -o objects.o convolve.o bessel.o utilities.o
```

(and possibly other flags on some platforms) after which the file may be dynamically loaded within S-PLUS by using

```
dyn.load("objects.o")
```

The result is an invisible character string vector giving the entry points of the routines loaded. (Note: multiple files can be specified directly to dyn.load, but they cannot then reference later files in the same load.)

Libraries can also be specified on the ld command, but on dynamically-loading systems these must be the *static* versions (such as /lib/libm.a). Do *not* include libc.

There is a COMPILE facility that handles most of the compilation details, so our convolve.c file was compiled and loaded as follows:

```
> !Splus COMPILE convolve.c
targets= convolve.o
make -f /usr/local/splus/newfun/lib/S_makefile  convolve.o
cc -c -I${SHOME-'Splus SHOME'}/include -O2   convolve.c
> print(dyn.load("convolve.o"))
[1] "_convolve"
```

Note that this will consult the Makefile (or makefile) in the current directory, which can be used to set or change the compiler flags. (Beware: despite its name, this is a make facility. Thus it will not recompile name.o if it is newer than its source file, and if both name.c and name.f are present, Splus COMPILE name.c will compile name.f !)

3. *Shared libraries* with dyn.load.shared (Sun Solaris 2.x, SGI and DEC Alpha). This is another form of incremental loading that is supported by some UNIX systems. The usage is

```
dyn.load.shared("./shlib.so")
```

where `shlib.so` is a shared library; the argument should be a path containing `"/"`. The object (`.o`) files in a shared library must be position-independent which may necessitate special compiler options. This is handled by the `SHLIB` script, so

```
Splus SHLIB -o name.so object1.o [object2.o ...]
```

will compile source code to produce the specified objects and then combine them into a shared library. Thus for the `convolve.c` example we could use

```
Splus SHLIB -o convolve.so convolve.c
```

4. *Augmented loading* with `dyn.load2` (SunOS4, HP-UX). This is similar to `dyn.load` except that it works by loading the extra routines into central memory, temporarily augmenting the symbol table, recording the locations of the new routines and finally discarding the augmented symbol table. The result is that each time new routines are loaded they cannot connect with previously loaded routines. `Splus COMPILE` can be used for compilation. It is necessary to tell `dyn.load2` how much space to reserve for the object by its `size` argument: the default ($10^5$ bytes) is usually sufficient, but if not the warning message will say how much is needed.

## S-PLUS 5.x on UNIX

This is the simplest case: there is only one method available, shared libraries, and the easiest way to include compiled code is to include the files in a call to `Splus5 CHAPTER`, for example

```
Splus5 CHAPTER convolve.c
Splus5 make
```

This will create a shared library called `S.so` in the chapter. Then the next time S-PLUS is started in that chapter, `S.so` will be loaded. Also, whenever a chapter (including a library) is attached the system looks to see if it contains a file `S.so` and if so will load it.

It is possible to use `dyn.open` and `dyn.close` to load or unload a shared library, but this is not normally necessary. Sometimes it easiest to use `dyn.open` to re-load the routines after re-compiling them, although calling `synchronize` on the chapter will unload and re-load `S.so`.

If `Splus5 CHAPTER` is called with no arguments it will create a `makefile` which will compile and link in all the C and FORTRAN (or Ratfor) files in the directory. If a `makefile` or `Makefile` already exists it will be amended as S-PLUS sees necessary.

The flags for compilation can be changed by setting `CFLAGS`, `FFLAGS` or `CXXFLAGS` as appropriate: this may well be necessary as up to 5.1 release 1 the default flags omitted optimization and (at least on Solaris and Linux) the flags needed for position-independent code.

## S-PLUS 3.x on Windows

Using compiled code needs either a Watcom compiler that is no longer available or a compiler than can produce 16-bit DLLs (and few of those are still available). If you have such a compiler, please consult the Programmer's Manual.

## S-PLUS 4.x and 2000 on Windows

The preferred compilers were Watcom 10.x, but these are no longer available.

1. *Static loading* can be used with the preferred Watcom compilers, only.

2. *Incremental loading* with dyn.load. This needs Watcom 10.x or 11.0 compilers, which are no longer on sale. The dyn.load function dynamically loads object files and makes the functions available to the version of S-PLUS presently running. There is a COMPILE batch script which can compile more than one source file, but they must either be all C or all FORTRAN. *A warning:* the batch script will not cope with filenames containing spaces, so if necessary use the short-names version of the pathname for SHOME.

   All the objects have to be loaded together on the call to dyn.load. Their order will be important, as they are loaded incrementally and only entry points in those routines already loaded can be resolved. A typical call (for our MASS library) is

   ```
   dyn.load(c("MASS.obj", "ppr.obj"))
   ```

   According to the Programmer's Guide, it is not possible to do source-code debugging of code loaded by dyn.load.

3. *Dynamic Link Libraries.* These are the equivalent under Windows of UNIX shared libraries, and are loaded using the function dll.load. Although there are standard specifications of DLLs, the details of how these are created and how their entries are called are highly compiler-specific, discussed further in Section A.4. The advantage of using DLLs is that, probably, your favourite compiler can be used, and high-quality free compilers can be. It is possible to do source-code debugging of DLLs loaded by dll.load (see page 148). It is relatively easy to mix C and FORTRAN code in a single DLL.

   Suppose that convolve.c has been made into convolve.dll. We can load the DLL and test that the symbol has been recognised by using the appropriate names in

   ```
   > dll.load("convolve.dll", call="cdecl", symbols="convolve")
   [1] 1
   > dll.load.list()
   [1] "convolve.dll" "slapi"
   > dll.symbol.list()
   [1] "S_api_get_message" "convolve"
   ```

(The argument `calling.convention` needs to be set appropriately. The options are `"stdcall"` and `"cdecl"`; `"cdecl"` is normally appropriate, but it is not the default.) Note that *all* the symbols to be used do need to be declared in a call to `dll.load`, and there is no mapping of symbol names as in `.Fortran` (which maps to uppercase on this platform).

A call to `dyn.load` that successfully loads the DLL returns `1`; subsequent calls can be used to declare further symbols and will return `0`. Negative return values indicate errors: use `?dll.load` for details (but we have found the precise error indication to be unreliable). The functions `dll.load.info`, `dll.symbol.info`, `is.dll.loaded` and `is.dll.symbol` can provide more specific information on which symbols are loaded.

We can test the code and unload the DLL by

```
> u <- rep(1, 5)
> u %+% u
[1] 1 2 3 4 5 4 3 2 1
> u %+% u %+% u
 [1]  1  3  6 10 15 18 19 18 15 10  6  3  1
> dll.unload("convolve.dll")
```

It is desirable to unload the DLL before loading a new version, and Windows will not permit the creation of a new version of the DLL while it is loaded.

## A.2   Procedures with R

### Under UNIX

R uses shared libraries, but the load command is called `dyn.load`, in contrast to S-PLUS. Shared libraries usually have extension `.so`, but some systems use `.sl`. As under S-PLUS, a shared library can be built by

```
R SHLIB convolve.c [...]
```

optionally using the `-o` flag to set the name of the library, which here will default to `convolve.so` or `convolve.sl` as appropriate. This can then be loaded by

```
dyn.load("./convolve.so")
```

Shared libraries can be generated automatically as part of the installation of an R package.

## Under Windows

R on Windows uses `dyn.load` for the loading of DLLs which are created in the same way as those for S-PLUS 4.x; indeed the same DLLs can be used provided they do not reference any variables nor functions in the engine. See Section A.4 for further details. The generation of DLLs is normally done as part of the compilation of a package, and this can be done via a standard `Makefile`; see Section 8.5. It is desirable to unload the DLL with `dyn.unload` before loading a new version, and Windows will not permit the creation of a new version of the DLL while it is loaded.

## A.3  Common concerns

Ideally, routines should be dynamically loaded only once, although the latest version loaded will be used,[1] which can be useful when developing code. The function `is.loaded` may be used to determine if some particular function has been loaded. The argument to `is.loaded` is the name of a function as a character string, but the name as it is known to the symbol table rather than as it may be known to the programmer. From the display on page 236 it can be seen that the C function `convolve` has been given the name `"_convolve"` on the system used (SunOS4). The function `symbol.C` may be used to find the symbol table name corresponding to any C function name. Similarly `symbol.For` can be used for FORTRAN subroutines.[2] (It is not safe to assume that FORTRAN names have a trailing underscore nor that they are distinct from C names; they share a name space on HP systems and on some Windows compilers.) For example consider extending our `%+%` function so that if the C function `convolve` is not available the file `convolve.o` is dynamically loaded first:

```
"%+%" <- function(a, b) {
  if(!is.loaded(symbol.C("convolve"))) dyn.load("convolve.o")
     . . . .
}
```

Other possibilities are to include the dynamic loading in the session startup function `.First`, and for libraries to use `.First.lib` (see page 195).

Under S-PLUS 5.x there is an alternative function `dyn.exists`

```
> is.loaded(symbol.C("convolve"))
[1] T
> dyn.exists(symbol.C("convolve")) != 0
[1] T
```

---

[1] except under Solaris with `dyn.load.shared`; see its on-line help for a workaround.

[2] There may be problems with the case which `symbol.For` does not cover: on most UNIX systems symbol table names are in lower case whatever case is used for the subroutine name in the FORTRAN code. Thus on Solaris subroutine `BDRppr` will have symbol table name `bdrppr_`.

## Using C++

The issues with C++ are to ensure that the C++ libraries are linked and that the C++ startup and terminate code is run. We have only used C++ with S-PLUS and R systems that used shared libraries and can thereby load the C++ libraries when they are loaded. On such systems (including S-PLUS 5.x and R on Solaris and Linux) all that is needed is to link the shared library using the C++ compiler as the linker.

## A.4   Writing Dynamic Link Libraries for Windows

The use of dynamic link libraries (DLLs) changed very considerably between S-PLUS versions 3.x and 4.x, both because 32-bit DLLs are required and since the S-PLUS interface has been changed. R for Windows uses DLLs that are built in a very similar way to those for S-PLUS 4.x and they may even be interchangeable. We only consider such 32-bit DLLs here.

### Generating the DLL

How to generate the DLL depends on the compiler. We will illustrate it by building a DLL of `convolve.c` (Figure 6.1 on page 126) using various compilers.

A 32-bit DLL may have a single main routine giving actions to be performed on loading and unloading. This will normally do nothing and may be omitted under most compilers/linkers, including those used here.

*Microsoft Visual C++*

Under Microsoft Visual C++ we can use the file `conVC.c` which differs only in the declaration of `convolve`

```
__declspec(dllexport) /* or LibExport  if <S.h> is included */
void convolve(double *a, long *na, double *b, long *nb,
              double *ab)
{
  int i, j, nab = *na + *nb - 1;

  for(i = 0; i < nab; i++)
    ab[i] = 0.0;
  for(i = 0; i < *na; i++)
    for(j = 0; j < *nb; j++)
      ab[i + j] += a[i] * b[j];
}
```

The simplest way to build a DLL is within the development environment, but it can also be built from the command line by

```
cl /MT /Ox /D WIN32  /c conVC.c
link /dll /out:convolve.dll conVC.obj
```

An alternative is to use `convolve.c` unchanged, but to declare the symbol on the link command by

```
link /dll /out:convolve.dll /export:convolve convolve.obj
```

or via a `.def` file.

If the code calls entry points in the S-PLUS engine, add

```
%S_HOME%\lib\Sqpe.lib
```

to the `link` command, having set S_HOME appropriately, or in the IDE add it to the libraries to be used. This import library is part of the optional development tools which can be added from the installation CD, if they have not already been installed. This library may not have been created for your version of Visual C++, but a suitable version can be created (in the current directory) by

```
lib /def:%S_HOME%\lib\Sqpe.def /machine:ix86 /out:Sqpe.lib
```

Note that it may not be obvious that code calls entry points as these can be used by macros included from `S.h` or further header files it includes.

If the code calls entry points in the R engine, create an import library from the file `R.exp` in the source-package distribution by

```
lib /def:R.exp /machine:ix86 /out:Rdll.lib
```

and add

```
Rdll.lib
```

to the `link` command or add it to the libraries to be used in the IDE.

*Watcom C++*

Using Watcom C++ 10.x and 11.0 the code needed is unchanged. We used a link file `conWAT.lnk`:

```
system nt_dll initinstance terminstance
export convolve
file convolve
name conWAT
```

and compiled and linked by

```
COMPILE convolve.c
wlink @conWAT
```

to generate `conWAT.dll`. Multiple object files can be combined into a single DLL by specifying them on the `file` line and listing all the symbols to be exported.

The appropriate import library for S-PLUS with the Watcom compilers is

```
%S_HOME%\lib\Sqpew.lib
```

in the optional development tools, or one can be built by

```
wlib Sqpew.lib %S_HOME%\cmd\Sqpe.dll
```

(which does not need the development tools installed).

*GNU compilers*

The `mingw32` port[3] of the GNU compilers (`gcc`, `g77` and so on) is the most appropriate one: the `cygwin` port can also be used with the `-mno-cygwin` flag. For a recent version of the compilers[4] building the DLL is easy:

```
gcc -O2 -c convolve.c
dllwrap -o convolve.dll --export-all-symbols convolve.o
```

It is possible to export symbols selectively by specifying them in a `.def` file. Perhaps the easiest way is generate a list and edit it, as in

```
gcc -O2 -c -I%S_HOME%/include big1.c big2.c ...
dlltool --export-all-symbols --output-def big.def big1.o ...
<edit big.def>
dllwrap -o big.dll --def big.def big1.o big2.o ...
```

An import library for S-PLUS can be built[5] by copying `Sqpe.dll` and `Sqpe.def` to the current directory and running[6]

```
dlltool --dllname Sqpe.dll --def Sqpe.def --output-lib libSqpe.a
```

Generally `gcc` follows Visual C++ conventions, so it may be useful to define `_MSC_VER` when using the S-PLUS header files.

A FORTRAN example may be helpful: file `testit.f` contains

```
      subroutine TESTIT(x, n, m)
      dimension x(n)
      do 10 i = 1, n
10      x(i) = x(i)**m
      end
```

which we compile and build a DLL from by

```
g77 -O2 -c testit.f
dllwrap -o testit.dll --export-all-symbols testit.o
```

To use this from S-PLUS we can use

```
dll.load("testit.dll", call="cdecl", symbols="testit_")
.C("testit_",as.single(1:5),as.integer(5),as.integer(-2))[[1]]
```

Note the use of `.C` not `.Fortran` and that the entry point is written in lowercase and given a trailing underscore. (Function `.Fortran` can be used, but the trailing underscore must still be added.)

The functions DBLEPR, INTPR and REALPR (page 133) in S-PLUS are written for the Watcom FORTRAN compiler and cannot be used directly with `g77`. We have written some interface code in C; see the on-line material for the book. Use this by

---

[3]from `ftp://ftp.xraylith.wisc.edu/pub/khan/gnu-win32/mingw32` at the time of writing.

[4]we used `gcc 2.95.2`.

[5]a couple of entries in `Sqpe.def` ending in ' . ' had to be removed in the version we tried.

[6]This may take several minutes at least, especially on Windows 9x.

```
gcc -c glue.c
g77 -O2 -c my.f
dlltool --export-all-symbols --output-def my.def my.o
dllwrap -o my.dll --def my.def my.o glue.o -L. -lsqpe
```

Those functions can be used in R without such trickery.

FORTRAN character strings cannot be used with S-PLUS with this compiler except via a C interface function: if writing one beware that they are not null-terminated.


## Checking exports

It is often helpful to check that the DLL contains the exports (and imports) that are expected. If Visual C++ is available one can use

```
dumpbin /exports /imports name.dll
```

Alternatively, a free program pedump.exe can be found on the Web, with syntax pedump -e -i name.dll.

This can be particularly helpful in finding out what transformations have been applied to Fortran symbols.


## Calling S-PLUS entry points

It is possible to use routines within the S-PLUS DLL by importing their symbols (from Sqpe.dll), but the details can be tricky. As Sqpe.dll was built using the Watcom compilers they present few problems, but other compilers may have different conventions for arguments and especially return values.

It is possible for both C and FORTRAN routines in a DLL to perform standard I/O. For C the include files mentioned on page 132 need to be included: for FORTRAN the routines DBLEPR, INTPR and REALPR are used (see page 133). Various entry-points will need to be declared as imports from Sqpe.dll at the link phase: the simplest way is to use an import library as described earlier under each compiler.

There are very many exported symbols in Sqpe.dll but only a few are documented (see Table 6.2 on page 129). Note that the S-PLUS header files are set up to work with Visual C++ as well as Watcom compilers. However, there are problems with return values from functions in Sqpe.dll (which use Watcom conventions), so it is not straightforward to use the random-number functions, for example. S-PLUS 2000 has macros S_DOUBLEVAL and S_FLOATVAL defined in compiler.h (included by S.h) that work around this for Visual C++, and the same macros can be used with gcc.

For example, we can write VCrnd.c:

```
#include <S.h>
LibExport void urand(long *m, double *p)
{
  int i;
  seed_in((long*)NULL);
  for (i = 0; i < *m; i++)
    p[i] = S_DOUBLEVAL(unif_rand());
  seed_out((long*)NULL);
}
```

and can compile and test this by

```
cl /MT /Ox /D "WIN32" /I %S_HOME%\include /c VCrnd.c
link /dll /out:VCrnd.dll VCrnd.obj %S_HOME%\lib\Sqpe.lib

> set.seed(123); runif(4)
[1] 0.8756982 0.5321866 0.6700785 0.9921576
> dll.load("VCrnd.dll", call="cdecl", symbols="urand")
[1] 1
> set.seed(123)
> .C("urand", as.integer(4), x=double(4))$x
[1] 0.8756982 0.5321866 0.6700785 0.9921576
```

Using gcc, we wrote a file VCrnd.def

```
LIBRARY VCrnd
EXPORTS
  urand
```

and built the DLL by

```
gcc -O2 -c -D_MSC_VER -I %S_HOME%/include VCrnd.c
dllwrap -o VCrnd.dll --def VCrnd.def VCrnd.o -L. -lSqpe
```

then used it in exactly the same way.

## Calling R entry points

Once again this is easiest if the DLL is built with the same compiler(s) used to build the R engine R.dll, the mingw32 versions of gcc and g77. These appear to use the same calling conventions as Visual C++, so using that also seems quite successful. Conversely, using a Watcom compiler is fraught with difficulties. The package-build sources for R for Windows are needed.

At the compile stage, the header file globalvar.h may need to be included, as this declares as imports all the global variables exported by R.dll. Do check that your code is not using any of the variable names defined there.

When linking, you will need an import library. The package-build sources will make one for gcc by make libR.a and we can build an import library R.lib for Visual C++ by lib /def:R.exp /out:Rdll.lib. We can illustrate this by generating a few random numbers with a DLL built by Visual C++ using file VCrndR.c:

```
__declspec(dllimport) double unif_rand();
__declspec(dllimport) void seed_in();
__declspec(dllimport) void seed_out();

__declspec(dllexport) void urand(int *m, double *p)
{
  int i;
  seed_in((long*)0L);
  for (i = 0; i < *m; i++) p[i] = unif_rand();
  seed_out((long*)0L);
}
```

We can compile and test this by

```
cl /MT /Ox /D "WIN32"  /c VCrndR.c
link /dll /out:VCrndR.dll VCrndR.obj Rdll.lib

> dyn.load("/path/to/VCrndR.dll")
> .Random.seed <- c(1, 1:3)
> .C("urand", as.integer(4), x=double(4))$x
[1] 0.5641106 0.2932388 0.6696743 0.9174765
```

which is the same answer as runif(4) starting from that seed.

## Using C++

All we need to do is to ensure that the correct libraries and C++ initialization code gets invoked. Suppose funs.cc contains your C++ code, and SR.cc is the wrapper with the linking functions declared as extern "C". Then to build a DLL that can be linked into either S-PLUS or R we can use

```
g++ -c funs.cc SR.cc
dlltool --export-all-symbols --output-def funs.def SR.o
dllwrap -o my.dll --def funs.def funs.o SR.o -lstdc++
```

with gcc and

```
cl /c funs.cpp SR.cpp
link /dll /out:my.dll /export:linksym funs.obj SR.obj
```

with Visual C++, which requires the .cpp extension. In each case the C++ startup code will be executed when the DLL is loaded, and the terminate code when it is unloaded, so static variables will be constructed and deconstructed correctly.

# Appendix B

# The Interactive Environment

The interactive environments can be customised in a number of ways to make the programmer's life a little easier.

## B.1 History and audit trails

### S-PLUS under UNIX

Unless disabled, S-PLUS keeps a compact record of all commands in a file .Audit in the current working database. This can be accessed in two ways under UNIX. The `history` command will retrieve commands (by default the last 10) and allow them to be edited and re-submitted, and `again` does so for the last command. Both can search for commands matching a pattern, their first argument. For example

```
history(max = 50, rev = F, evaluate = F)
```

will list the last up to 50 commands in chronological order, and

```
again("lda", ed = T)
```

allows the last command containing the string "lda" to be edited and then submitted.

The command `Splus AUDIT` used from the UNIX operating system prompt allows a much more detailed investigation of the audit trail. It has a cryptic command language detailed in its help page. For details of the power of this facility, see Becker & Chambers (1988).

The audit trail can grow large. The option `audit.size`, default 0.5Mb, if reached triggers a warning at the beginning of a session. Run (at the UNIX prompt)

```
Splus TRUNC_AUDIT n
```

at any time (outside S) to truncate the file to about $n$ bytes (default 100000). Auditing can be disabled by ensuring that the audit file is not writable, for example by running the following commands at the UNIX prompt

**Figure B.1**: The history buttons in S-PLUS 2000. On the left is the button that brings up the History script window, then the button for a commands window (depressed) and on right the button for the Commands History window.

```
touch .Data/.Audit
chmod o-w .Data/.Audit
```

which creates a file if one does not exist, then marks it as read-only.

### S-PLUS under Windows

An audit file `_Audit` may be created under Windows, but there are no equivalents of `history`, `again` and `Splus AUDIT`. The default is to disable auditing by setting the environment variable S_NOAUDIT=X. To re-enable auditing, set S_NOAUDIT= in the command used to start S-PLUS (usually in the properties of a shortcut). The Windows command to truncate the audit file is TRUNC_AU n.

S-PLUS for Windows has an alternative command-history mechanism. There is a Commands History button (see Figure B.1) on the main toolbar. This brings up a dialog box within which commands can be reviewed, selected, edited and executed. These commands can be saved in a file from that window's File menu. Setting the environment variable

```
S_CMDFILE=+history.q
```

causes the command history to automatically be appended to the file[1] history.q in the current directory, and this file will be used in subsequent sessions to initialize command-line editing (so using the up arrow can scroll back into previous sessions) and the Commands History window. Setting both S_NOAUDIT= and S_CMDFILE=+_AUDIT turns on auditing and will use the audit file to initialize and record the commands history. Auditing stores much more information, but the Commands History window will only show the commands. The number of lines saved for command-line recall or Commands History can be limited by setting the S_CMDSAVE variable to the desired (numeric) limit.

It also has a History window, a script window which will show both S commands and operations performed in the GUI during the current session. This too has a button on the main toolbar (see Figure B.1).

### R under UNIX

Most UNIX versions of R are compiled to use the GNU `readline` library, which provides a commands history mechanism unless R is invoked with the `--no-readline` option. The commands are saved to a file `.Rhistory` in the

---

[1]creating it if it does not exist.

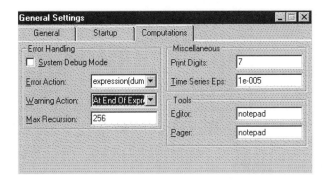

**Figure B.2**: Part of the Computations tab of the General Settings ... dialog box from the Options menu in S-PLUS 2000.

start-up directory, provided the workspace is saved, and reloaded into the commands history for the next session. File .Rhistory is a plain text file, and can be edited or even replaced to initialize the commands history in the following session.

### R under Windows

The GUI Windows front-end for R saves a record of commands in the file .Rhistory in the working directory, and reloads it at the start of the next session. Unlike UNIX versions, it does this whether or not the workspace is saved.

## B.2 Options

The options function can be used to set defaults for many aspects of the engine in use. Some of these can be set from the Options menu in S-PLUS for Windows versions: see Figure B.2.

The function options accesses or changes the dataset .Options, which can also be manipulated directly. Its exact contents will differ between operating systems and engines, but part of one example (S-PLUS 5.1) is

```
> unlist(options())
      echo prompt continue width length        keep check digits
   "FALSE"    "> "      "+ "   "80"   "48" "function"   "0"    "7"
      memory object.size audit.size          error    show
"2147483647"        "Inf"    "500000" "dump.calls" "TRUE"
   compact scrap free warn editor expressions reference
"100000" "500"   "1"  "0"   "vi"        "256"       "1"
contrasts.factor contrasts.ordered verify completions
   "contr.helmert"      "contr.poly"     "1"     "FALSE"
conflicts.ok indentation  ts.eps sequence.tol time.month.name1
      "FALSE"        "\t" "1e-05"      "1e-06"        "January"
```

```
     . . . .
pager help.pager help.browser
"less"    "slynx"    "netscape"
```

Other options (such as `gui` in S-PLUS 3.4) are by default unset. Calling
`options` with no argument or a character vector argument returns a list of the
current settings of all options, or those specified. Calling it with one or more
`name=value` pairs resets the values of component `name`, or sets it if it was unset.
For example,

```
> options("width")
$width:
[1] 80
> options(width=65)
> options(c("length", "width"))
$length:
[1] 48
$width:
[1] 65
```

The meanings of the more commonly used options are given in Table B.1. In the
new S engine changes made by calling `options` within a function are local to
the current top-level expression unless the argument `TEMPORARY=F` is set.

S-PLUS has a number of other options which can be used to tune its performance, but these are for use by gurus.

There are a number of options that can be set by environment variables, some
of which we saw in the previous section. Details differ by engine, but some useful
ones for R are

R_LIBS        where to search for libraries
R_VSIZE       the size of the heap (usually followed by M for megabytes).
R_NSIZE       the number of 'cons cells'
R_PROFILE     where the .Rprofile file is (see page 252).

These can be set in many ways, perhaps most conveniently in a file .Renviron
in the current or home directory, for example

```
R_VSIZE=10M
R_NSIZE=400k
R_LIBS=/ext/R/library
export R_VSIZE R_NSIZE R_LIBS
```

This also works in the Windows version of R, although the last line is not needed.

## B.3   Session startup and finishing functions

One way to customize the environment is to run some S code at the start or end
of every session. Under S-PLUS the easiest way to do so is to use a .First

**Table B.1**: Some commonly used options to customize the S environment.

| | |
|---|---|
| editor | The default text editor for `ed` and `fix`. |
| pager | The default pager for `help`, `page` .... |
| width | The page width, in characters. Not always respected. |
| length | The page length, in lines. Used to split up listings of large objects, repeating column headings. Set this to a very large value to suppress this. |
| digits | Number of significant digits to use in printing. Set this to 17 for full precision. |
| echo | Logical for whether expressions are echoed before being evaluated. Useful when reading commands from a file. (Exists in R, but is not respected.) |
| prompt | The primary prompt. |
| continue | The command continuation prompt. |
| error | Function called to handle errors. See Section 8.2. |
| warn | The level of strictness in handling warnings. The default, 0, collects them; 1 reports them immediately and 2 makes any warning an error condition. Negative numbers cause warnings to be ignored. |
| memory | (S-PLUS) The maximum memory (in bytes) that can be allocated. It can be helpful to reduce this when experimenting with memory-intensive code. |
| object.size | (S-PLUS) The maximum size (in bytes) of any object. The default in the old S engine is quite low and can usefully be increased. |
| expressions | The maximum depth of nesting expressions. |
| ts.eps | A tolerance used by some time series functions in determining whether a time coincides with an observation time. |
| conflicts.ok | (new S engine) warn about masked functions when a database is attached. |
| indentation | (new S engine) indentation in code listings, for example, `indentation=" "`. |

or .Last function, which are run during the startup and (normal) shutdown sequences. Recent Windows versions also allow a function .guiFirst which is run much later in the sequence than .First, when the GUI has been initialized, so can be used to load menus, for example. All of these should be functions without arguments: they are called as .First() and so on.

In fact any database, for example a library, can contain a .Last function which will be run when the database is detached.

The S_FIRST environment variable may contain code that is executed at startup: if it is set, .First will not be called. (This provides a way to start S-PLUS if the .First function is erroneous, but always test a .First function before quitting the session.)

.First and .Last functions can also be used in R. However, unless there is one in a workspace that is restored, there will not be a .First function available at startup. It is usually more convenient to put startup code in a .Rprofile file. Unless --no-init-file was specified, R looks for .Rprofile first in the current directory then in the user's home directory and sources the first one it finds. This is done before any .First is called.

The new S engine has the further option of a .S.init file, the analogue of R's .Rprofile. This should contain S code to be source-ed, and is looked for first in the current directory then the user's ~/MySwork directory. This is run before .First. Even earlier in the sequence the file .S.chapters (from the same places) is used to attach S chapters or libraries (listed one per line in the file).

Actually, all the new S engine startup actions are controlled by the file $SHOME/S.init which can be replaced but is normally the S expression

```
{
standardInitialize()
for(where in c(".", getenv("S_MYSWORK")))
    if(where != "" &&
      canOpen(paste(where, ".S.init", sep="/"), "r")) {
        source(paste(where,".S.init",sep="/"))
        break
    }
.doAction(".First.Sys")
if(getenv("S_FIRST") == "") .doAction(".First")
else eval(parse(text=getenv("S_FIRST")))
}
```

This shows precisely how the startup actions are sequenced (.First.Sys is a system function that experts can change). The function standardInitialize is essentially

```
standardChapters()
standardWorkingData(T)
setAudit()
```

# Appendix C

# BATCH Operation

## C.1 S-PLUS

It is sometimes desirable to run an S-PLUS job non-interactively. On a UNIX machine under `csh` we might use

```
Splus < infile >& outfile &
```

but unless `options(echo=T)` has been set the input commands will not be echoed to the output file. (We use `>&` to redirect prompts and errors to the output file.) The `Splus BATCH` command automates this process, so

```
Splus BATCH infile outfile
```

sets `options(echo=T)` and then reads `infile` and redirects all output to `outfile`. The command is implemented using `nohup`, the precise effect of which varies with UNIX platform; on BSD systems (such as SunOS4) it reduces the priority of the process somewhat, and on all systems it blocks some software signals to ensure that the S-PLUS job continues to run when the shell terminates.

Although the input is echoed, under 3.x ends-of-lines are often changed to semi-colons within braced expressions, so the outfile can be hard to read.

Under S-PLUS 5.x the behaviour is a little different: prompts are not echoed when input or output appears to be from or to a file or pipe. If you want to emulate a terminal interface, the only way we know to do this is to use pseudo-ttys, which S-PLUS regards as terminals. Line feeds are inserted in the output more frequently than under 3.x.

### Windows

S-PLUS 4.x and 2000 provide a command-line interface program `sqpe.exe` whose input and output can be redirected and which can be used from an MS-DOS or command-prompt window. Both prompts and error messages are written to the output stream, but commands are not echoed unless `options(echo=T)` is set. The prompts will not appear in the output file unless the environment variable `ALWAYS_PROMPT` is set. Before using `sqpe` you may need to ensure that the environment variable `SHOME` is set: an example of its use (in an MS-DOS window) is

```
set ALWAYS_PROMPT=T
set SHOME=C:\Program Files\splus45
%SHOME%\cmd\sqpe < infile > outfile
```

There is also a BATCH switch with syntax

```
splus [S_PROJ=dir] /BATCH infile [outfile [errfile]]
```

where error messages are sent to `outfile` if `errfile` is not supplied. Setting S_PROJ is needed to change the working directory _Data from its default. The input commands are not echoed (use `options(echo=T)`) and the prompts will not appear in the output file. The files are specified relative to the working directory.

## C.2  R

UNIX versions of R have a BATCH command: use

```
R BATCH [options] infile [outfile]
```

If `outfile` is missing it defaults to *infile*.Rout, stripping any .R extension. Unlike Splus BATCH this does not run the job in the background; use shell facilities (usually ending the line by &) to do this. Whenever R detects that input is coming from a file, the input is automatically echoed to the output: `options(echo=TRUE)` is accepted but has no effect.

The Windows version has a front-end `rterm.exe` that can be run from a shell in a terminal window, and can be called from a batch (.bat or .cmd) file or shell script to do equivalent things to R BATCH.

# References

Numbers in brackets [ ] are page references to citations.

Abramowitz, M. and Stegun, I. A. (1965) *Handbook of Mathematical Functions with Formulas, Graphs and Mathematical Tables*. New York: Dover. [50]

Atkinson, A. C. (1985) *Plots, Transformations and Regression*. Oxford: Oxford University Press. [161]

Becker, R. A. (1994) A brief history of S. In *Computational Statistics: Papers Collected on the Occasion of the 25th Conference on Statistical Computing at Schloss Reisenburg*, eds P. Dirschedl and R. Osterman, pp. 81–110. Heidelberg: Physica-Verlag. [1, 2]

Becker, R. A. and Chambers, J. M. (1984) *S. An Interactive Environment for Data Analysis and Graphics*. Monterey: Wadsworth and Brooks/Cole. [2]

Becker, R. A. and Chambers, J. M. (1985) *Extending the S System*. Monterey: Wadsworth and Brooks/Cole. [2]

Becker, R. A. and Chambers, J. M. (1988) Auditing of data analyses. *SIAM Journal of Scientific and Statistical Computing* **9**, 747–760. [247]

Becker, R. A., Chambers, J. M. and Wilks, A. R. (1988) *The NEW S Language*. New York: Chapman & Hall. (Formerly Monterey: Wadsworth and Brooks/Cole.). [2, 3, 61, 66, 125, 139]

Breiman, L. (1996) Bagging predictors. *Machine Learning* **24**, 123–140. [175]

Chambers, J. M. (1998) *Programming with Data. A Guide to the S Language*. New York: Springer-Verlag. [1, 2, 61, 99, 106, 109, 119, 125, 135, 141, 180]

Chambers, J. M. and Hastie, T. J. eds (1992) *Statistical Models in S*. New York: Chapman & Hall. (Formerly Monterey: Wadsworth and Brooks/Cole.). [2, 79]

Cox, D. R. and Hinkley, D. V. (1974) *Theoretical Statistics*. Chapman & Hall. [163]

Davison, A. C. and Hinkley, D. V. (1997) *Bootstrap Methods and Their Application*. Cambridge: Cambridge University Press. [172]

Durbin, R., Eddy, S., Krogh, A. and Mitchison, G. (1998) *Biological Sequence Analysis. Probabilistic models of proteins and nucleic acids*. Cambridge: Cambridge University Press. [176, 177]

Everitt, B. S. (1994) *A Handbook of Statistical Analyses using S-Plus*. London: Chapman & Hall. [161]

Freund, Y. (1990) Boosting a weak learning algorithm by majority. In *Proceedings of the Third Workshop on Computational Learning Theory*, pp. 202–216. Morgan Kaufmann. [175]

Freund, Y. (1995) Boosting a weak learning algorithm by majority. *Information and Computation* **121**, 256–285. [175]

Freund, Y. and Schapire, R. E. (1995) A decision-theoretic generalization of on-line learning and an application to boosting. In *Proceedings of the Second European Conference
on Computational Learning Theory*, pp. 23–37. Springer-Verlag. [175]

Freund, Y. and Schapire, R. E. (1996a) Experiments with a new boosting algorithm. In *Proceedings of the Thirteenth International Conference on Machine Learning*, pp. 148–
156. [175]

Freund, Y. and Schapire, R. E. (1996b) Game theory, on-line prediction and boosting. In
*Proceedings of the Ninth Annual Conference on Computational Learning Theory*, pp.
325–332. [175]

Gentle, J. E. (1998) *Numerical Linear Algebra for Applications in Statistics*. New York:
Springer-Verlag. [29]

Ihaka, R. and Gentleman, R. (1996) R: A language for data analysis and graphics. *Journal
of Computational and Graphical Statistics* **5**, 299–314. [3]

Jackson, M. A. (1975) *Principles of Program Design*. London: Academic Press. [172]

Kalbfleisch, J. D. and Prentice, R. L. (1980) *The Statistical Analysis of Failure Time Data*.
New York: John Wiley and Sons. [216]

Knuth, D. E. (1968) *The Art of Computer Programming, Volume 1: Fundamental Algorithms*. Reading, MA: Addison-Wesley. [18]

Lange, K. (1999) *Numerical Analysis for Statisticians*. New York: Springer-Verlag. [176]

Ripley, B. D. (1996) *Pattern Recognition and Neural Networks*. Cambridge: Cambridge
University Press. [173]

Schapire, R. E. (1990) The strength of weak learnability. *Machine Learning* **5**, 197–227.
[175]

Schwartz, R. L., Olsen, E. and Christiansen, T. (1997) *Learning Perl on Win32 Systems*.
Sebastopol, CA: O'Reilly & Associates. [226]

Spector, P. (1994) *An Introduction to S and S-Plus*. Belmont, CA: Duxbury. [3]

Thisted, R. A. (1988) *Elements of Statistical Computing. Numerical Computation*. New
York: Chapman & Hall. [29]

Venables, W. N. and Ripley, B. D. (1999) *Modern Applied Statistics with S-PLUS*. Third
Edition. New York: Springer-Verlag. (First edition 1994, second edition 1997.). [v, 1,
3, 5, 29, 31, 34, 35, 37, 58, 60, 71, 72, 82, 83, 124, 153, 161, 171, 173, 175, 176, 178,
206, 216]

Viterbi, A. J. (1967) Error bounds for convolutional codes and an asymptotically optimal
decoding algorithm. *IEEE Transactions on Information Theory* **13**, 260–269. [176]

Waterman, M. S. (1995) *Introduction to Computational Biology. Maps, sequences and
genomes*. London: Chapman & Hall. [176]

# Index

Entries in this font are names of S objects. Page numbers in **bold** are to the most comprehensive treatment of the topic.